ROBOTICS

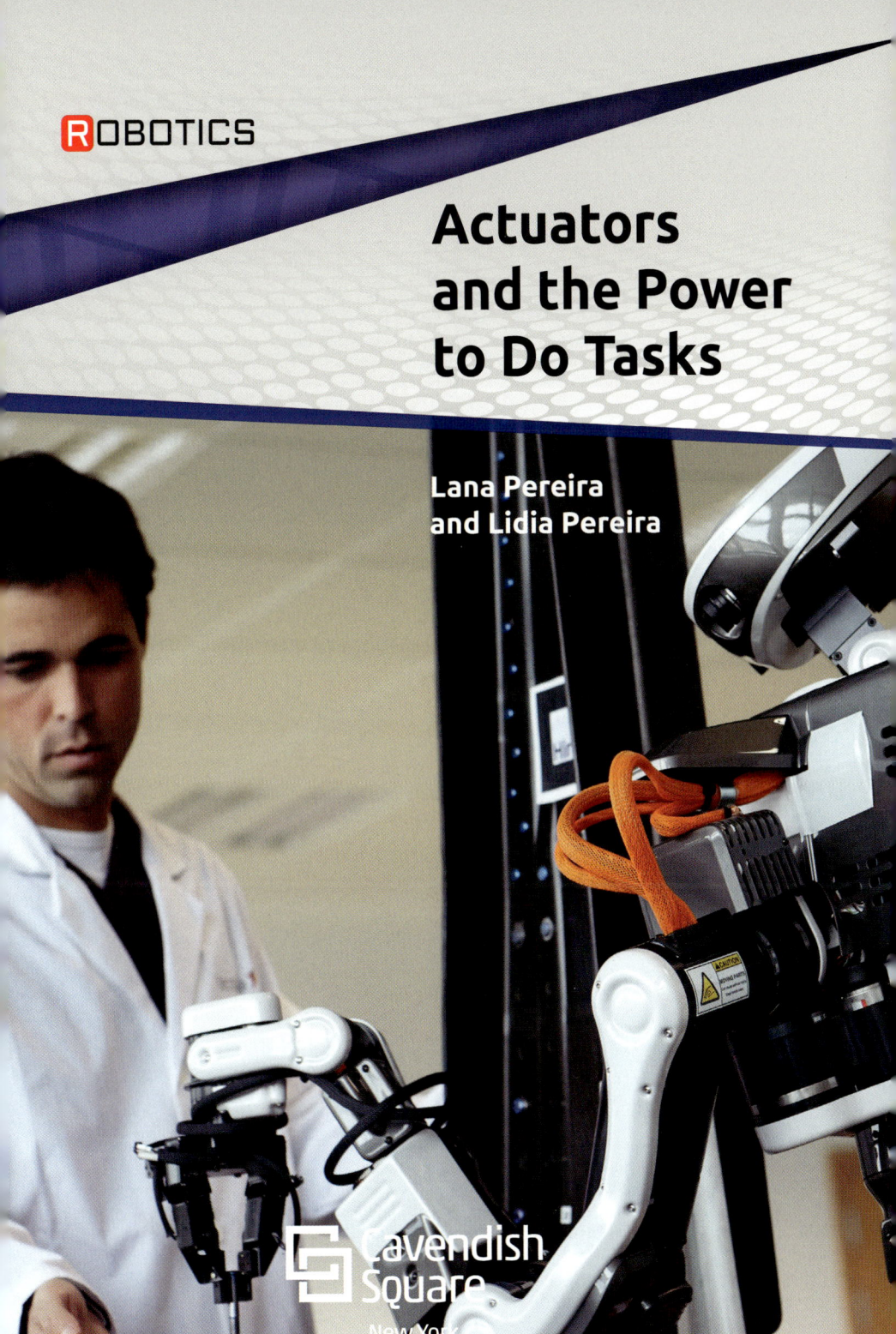

Actuators and the Power to Do Tasks

Lana Pereira
and Lidia Pereira

Cavendish Square
New York

Acknowledgments: The editors would like to thank John Long at Vassar College, Timothy Friez at Carnegie Mellon's Robotics Institute, Mishah U. Salman at Stevens Institute of Technology, and Suman Sabastin, a teacher in New York City, for their help in developing this series on robotics.

Published in 2017 by Cavendish Square Publishing, LLC
243 5th Avenue, Suite 136, New York, NY 10016

Copyright © 2017 by Cavendish Square Publishing, LLC

First Edition

No part of this publication may be reproduced, stored in a retrieval system, or transmitted in any form or by any means—electronic, mechanical, photocopying, recording, or otherwise—without the prior permission of the copyright owner. Request for permission should be addressed to Permissions, Cavendish Square Publishing, 243 5th Avenue, Suite 136, New York, NY 10016. Tel (877) 980-4450; fax (877) 980-4454.

Website: cavendishsq.com

This publication represents the opinions and views of the author based on his or her personal experience, knowledge, and research. The information in this book serves as a general guide only. The author and publisher have used their best efforts in preparing this book and disclaim liability rising directly or indirectly from the use and application of this book.

CPSIA Compliance Information: Batch #CW17CSQ

All websites were available and accurate when this book was sent to press.

Library of Congress Cataloging-in-Publication Data

Names: Pereira, Lana, author. | Pereira, Lidia, author.
Title: Actuators and the power to do tasks / Lana Pereira and Lidia Pereira.
Description: New York : Cavendish Square Publishing, [2017] | Series: Robotics |
Includes bibliographical references and index.
Identifiers: LCCN 2016029407 (print) | LCCN 2016036241 (ebook) |
ISBN 9781502619402 (library bound : alk. paper) | ISBN 9781502619419 (ebook)
Subjects: LCSH: Actuators--Juvenile literature. | Robotics--Juvenile literature. | Robots--Juvenile literature.
Classification: LCC TJ223.A25 P47 2017 (print) | LCC TJ223.A25 (ebook) | DDC 629.8/92--dc23
LC record available at https://lccn.loc.gov/2016029407

Editorial Director: David McNamara
Editor: Fletcher Doyle
Copy Editor: Nathan Heidelberger
Associate Art Director: Amy Greenan
Designer: Alan Sliwinski
Production Coordinator: Karol Szymczuk
Photo Research: J8 Media

The photographs in this book are used by permission and through the courtesy of: Cover Javier Larrea/Getty Iamges; p. 4 GOLFX/Shutterstock.com; pp. 8, 39 Designua/Shutterstock.com; pp. 11, 12, 20, 23, 27, 28, 33, 46, 47, 49, 50, 51, 52, 54, 55, 63, 64, 68, 83, 84, 94, 96, 103 Lana Pereira and Lidia Pereira; p. 14 David Madison/Getty Images; p. 32 Alfonso Gonzalez/Public Domain/File:Solenoid Valve.png/Wikimedia Commons; p. 36 Nick Ut/AP Images; p. 43 Leo Mason/Corbis/Getty Images; p. 48 Courtesy Parallax; p. 58 Getty Images; p. 61 Pavel L Photo and Video; p. 66 Hero Images/Getty Images; p. 75 TOSHIFUMI KITAMURA/AFP/Getty Images; p. 79 Sheila Fitzgerald/Shutterstock.com; p 89 NASA/File:S129e009180.jpg/Wikimedia Commons; p. 92 Rodrigo Espiritu/EyeEm/Getty Images; p. 100 Joyfuldesigns/Shutterstock.com; p. 112 David Paul Morris/Bloomberg/Getty Images.

Contents

Chapter 1: Get Moving   5

Chapter 2: Right Tool for the Job   15

Chapter 3: Accomplishing Your Tasks   37

Chapter 4: Help, My Robot Can't Get Up   67

Chapter 5: Moving Forward   93

Glossary   114

Further Information   120

Index   125

About the Author   128

Optimus Prime of the Transformers needs actuators to generate his movements.

1 Get Moving

How did the robot cross the road? With its wheels. How does a robot move his wheels? With its **actuators**. Actuators are responsible for creating movement. Whether it's turning a fan or making a robot walk, actuators are the things making the movement. Grab your pencil, and lift it—you were able to do this because your muscles stretched and contracted, powered by energy from your food. This caused your fingers and arm to move. The muscles behaved in the same way that a motor does in a robot: they have the ability to turn energy into motion. When you are riding your bike or pushing a skateboard, you are the actuator, the source of motion. Our energy comes from food, but a robot's energy can be derived from many different things. Some scientists are even working on unusual robots that derive their energy by breaking down dead insects, essentially by "eating" them. Most robots, especially those that you use at school, use electricity from batteries to power their actuators.

All robots, from **forklifts** to Optimus Prime, use actuators to move. So what is an actuator? In the simplest terms, an actuator is anything that takes energy and converts it to movement or force. Most often this energy comes from a battery, but it can come from anywhere. The actuators we usually use require electrical energy from batteries or a wall outlet. If you use a **pneumatic** system, then the energy is stored in the form of potential energy from pressurized gases. An internal combustion engine powers your car's actuators. In some cases, such as with certain **piezoelectrics** (electricity generated in certain materials by pressure), the energy can be as simple as heat. Actuators control a system, but they cannot "see" what they control. When garbage trucks pick up large bins, the actuators creating the motion can't tell if the bin falls off or something gets stuck. Most actuators provide no feedback.

There are many types of actuators, depending on the robot's job. Some of them are present in your everyday life. Your cell phone uses a small **DC motor** with an unbalanced weight to create vibrations. The hard drive in your desktop computer uses an actuator to spin the platters when you access information. Automatic garage door openers use a larger motor to lift the door. Even if the items you are using do not have actuators in them, there is a very good possibility they were manufactured using them. Factory robots, such as those that assemble cars, often use **servo motors** to control their movement. Servo motors provide feedback about their position, making them a good choice for factory robots that need to be **precise**.

Other actuators need force rather than precise adjustment, like the rollers in a pitching machine. When we mention factory robots, you likely think of robots painting cars or putting them together in factories, if you think of them at all. However, they can fulfill a huge range of tasks, from screwing furniture together to assembling a watch. Actuators like those we talk about in this book are also very common in machining, often used to cut metal or carve items like baseball bats. However, there is one way in which factory robots are different from the ones you will be building. Factory robots only fulfil a small task, over and over, as part of a larger **system**. The robots you will be building are more like a full factory than any one of those robots.

A robot alone can be thought of as a system. A system is defined as any group of things working together to form a unified whole. While you must understand each part of your robot as fully as possible, it is doubly important to understand how the parts interact to form the unified whole we call a robot. If you look at the example of a garbage truck again, you can clearly identify many different components in that system. You have the component that is the vehicle, the driver who operates the truck, and the actuator system that provides the heavy lifting capabilities. Each of these is a complex system of its own, but we tend to regard them as a single thing, a garbage truck. Each part needs to be able to accomplish its clearly defined tasks without interfering with the rest of the system. In this context, an actuator's task is to produce the robot's primary **output**, movement. Each actuator has a task to complete when it is asked to, whether it is moving the **drivetrain**, grabbing something, or lifting it. If an actuator does not operate when it is asked to, or operates when it is not asked to, that actuator is not doing its job. There are many reasons that can happen. Understanding how your actuators work is the first step to being able to use them well. Even the actuator itself is a complex system when you take a look inside it.

A Look Inside

The main components in DC motors are the shaft, the coils that are wrapped around the shaft, and the permanent magnet that surrounds the whole. The shaft and the coils wrapped around it are an example of a **simple machine**, a wheel and axle. Since they are functioning as a single piece and the coils

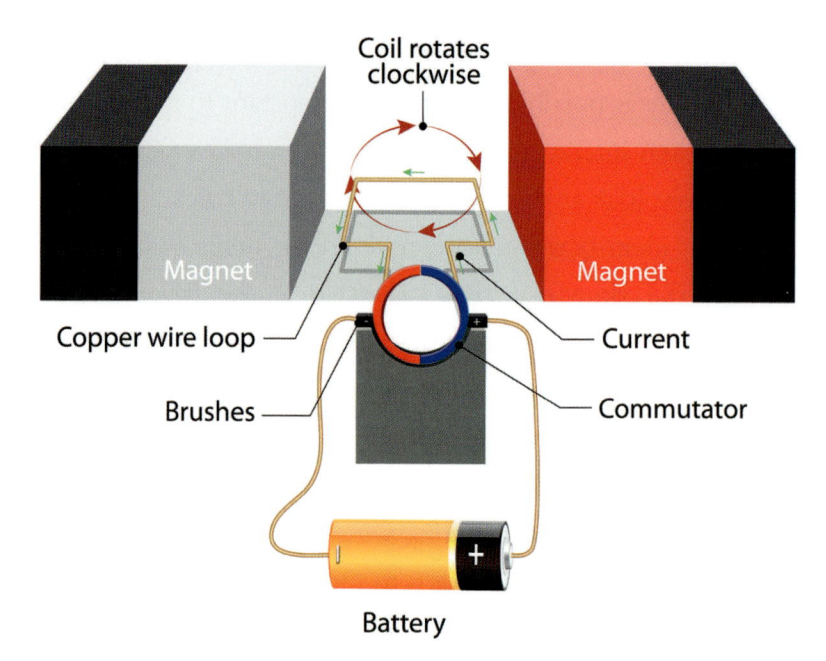

This is a diagram of a small DC motor.

are wider than the shaft, our ability to move the coils translates to **mechanical advantage** in moving the shaft. When current is applied to the coils inside the motor, it generates a **magnetic field**. The coil assembly is an **electromagnet**, which is anything that generates a magnetic field when a current is run through it. The electromagnet then moves to align itself with the permanent magnet in the casing, because magnetic fields do not like to be misaligned. As it rotates, the polarity of the field is reversed by switching the direction of current in the magnet. As long as we maintain the misalignment of magnetic fields, the electromagnet will move to align, causing the axle to rotate. The stronger the magnets, the more force the motor will supply. For any of this to happen, the actuator has to be supplied with power. We already know that the power will be coming from our battery. But in a larger system, we need a way to tell power where to go. This is done with a controller, like the EV3 Brick, or VEX's microcontroller. These are the brains of the operation—the big boss, where actuators are just henchmen. Actuators always need to be told what to do. These brains are what will be controlling the amount of power that goes to your actuator, which in turn affects what your motor will do. There are some motors that have chips in their casings. These are almost always what is best called a daughterboard. They hold components and communicate with the primary motherboard, but they cannot operate alone. The motherboard in our robots is the microcontroller, brick, or a similar "brain."

    We can't forget that the system above only applies to one type of actuator. There are **AC motors**, **stepper motors**, pneumatics, and even piezoelectric actuators that produce movements so precise that we must measure them in nanometers. There is a wonderful world of options out there. Even today, people are discovering new ways to produce movement. Some of the new designs include robots the size of postage stamps that can "walk" despite being 1 millimeter (0.04 inches) thick. But as with all new and exciting fields, you have to walk before you can run. Or in this case, roll before you can walk. We will have a much more narrow focus.

Common Systems

We will be discussing two different systems throughout this book. They have many advantages: both are well designed and well supported, and tailored to learning functional robotics and engineering. They are both inclusive kits, including all parts of many potential robots. The one more of you should be familiar with is LEGO Mindstorms, a modular system that is compatible with many types of LEGO. The latest revision of Mindstorms is the EV3, so we will focus on that over the NXT, the previous but very similar version. There are two different sizes of motors in EV3, the large and the medium, instead of NXT's one-size-fits-all system. EV3's large motor is fairly similar to the NXT's in **power** and **torque**. The medium motor is not suited to heavy loads and can operate at higher revolutions per minute (rpm) than the large. It is much more suited to precision applications that need speedy response times. We can look inside these just as we looked at the diagram of the DC motor.

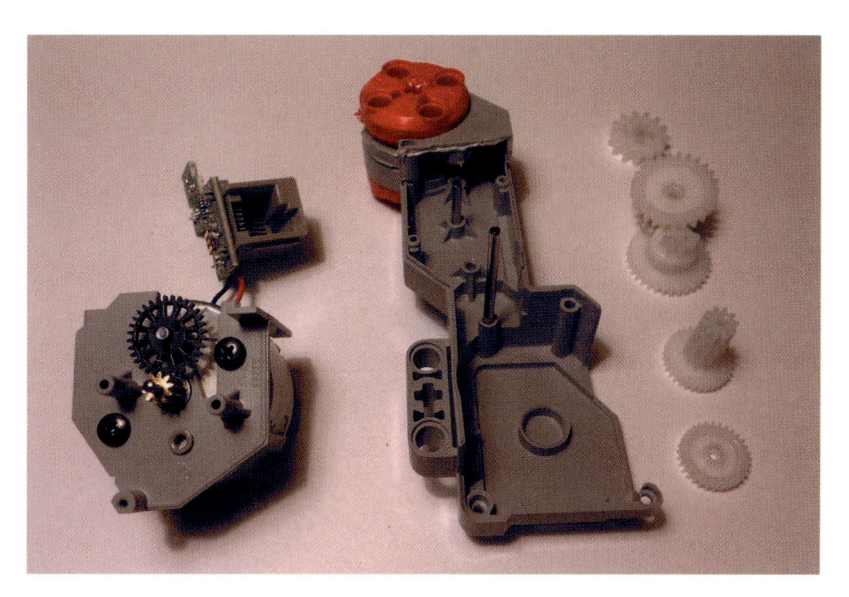

An opened EV3 large motor, showing the gears making up the gear train.

The EV3 motor (*above*) contains a series of **gears** that transmit power from the DC motor inside to the EV3's motor hub, in red. There is also a small, yellow tab soldered to the small board, which acts as a fuse, preventing the motor itself from burning out with a current overload. The wheels in the main compartment link to the **encoder** module, also on the PCB (printed circuit board). Notice the **pinion** in the main compartment. That is the **driving gear** that transfers motion from the axle of the motor to the gear train. When you're building your robot, you will be adding your own pinion gear.

The VEX system provides several options for actuators. If you know VEX, you very likely have already used several of them. VEX's "standard" actuator is a **rotational motor** with sufficient power to build drivetrains and various types of manipulators. The most common model is the VEX 2-Wire

Get Moving 11

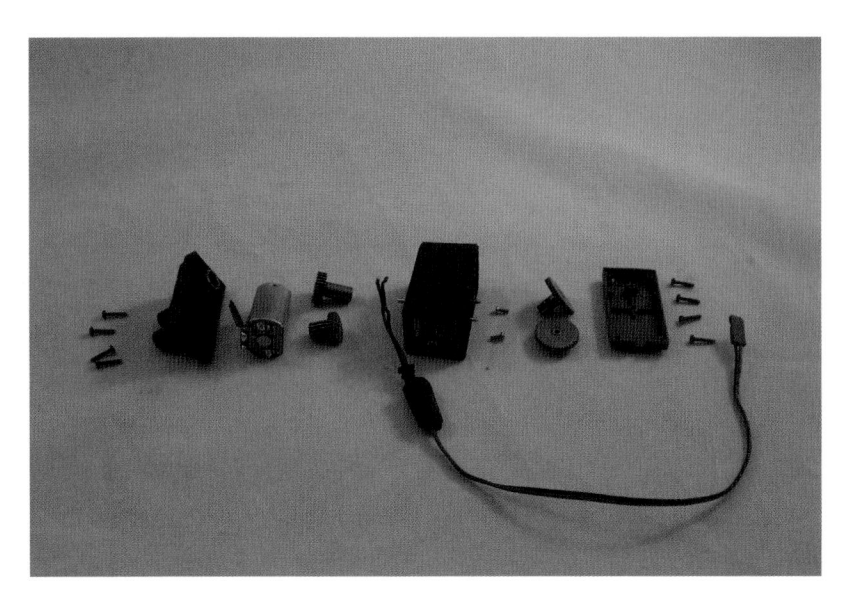

The component parts of a VEX motor.

Motor 393. You're actually somewhat likely to see the inside of the VEX motor yourself, because it contains a set of gears that can be swapped out to provide either higher torque or higher **speed**, depending on your goals.

Although the DC motors in the VEX and LEGO actuators are very differently shaped, they function essentially the same way. Actuators of any sort come in all shapes and sizes. The VEX system uses a very similar temperature fuse that prevents high currents from being pushed through the motor. Notice that in this motor there is no encoder, just the fuse and a PCB that manages a connector. The gears, which connect to the motor shaft, are the gears that can be swapped to gear torque up or down internally. Those pictured above are the standard gears, capable of about 100 rpm and up to almost 1.6 newton meters of torque. VEX also provides a servo motor, which uses an encoder to determine how far it has moved. The VEX servo, like many other servo motors, cannot complete a full rotation. Its approximate range of rotation is about 100 degrees. It also provides less torque than the 393 motor, since it is not likely to be used to power drivetrains or other heavy applications.

Now you have met the main characters. By the time you reach the end of this book, you will be able to build a robot that can boogie. In just a paragraph or two, you'll be reading about the most common actuators for robotics applications. Somewhere in here, we'll mention how important it is to wear safety glasses. We'll talk about how you can use actuators and how to choose a motor that's suited for your purpose in chapter 3. Chapter 4 shares how to care for and troubleshoot a sick or misbehaving robot. Lastly, we'll talk about what to do once you close the book—and where you can go from there. It's like riding a bike—you'll never forget how, but first you have to get up and go. So let's get actuating!

This pneumatic lug wrench uses rotational motion with a lot of torque to change a race car's tire quickly.

2 Right Tool for the Job

When you design a robot, you will need actuators in order to accomplish work in the form of movement. The actuator that you choose might vary depending on the movement, or output, you wish your robot to make. There is an enormous range of actuators available, powering everything from ATMs to electric vehicles. The choice for school projects is comparatively limited. Even so, motor selection can be of great importance. We mentioned that actuators control a system. It's important to understand that they have no brain, however, and cannot do thinking of their own. They control the system, but you control them. The motor is in charge the same way your remote is in charge of your television. It can control the TV, but only when it has **input** from you. When you are choosing your motor, there are many factors you need to consider. For most school projects, you will likely rely on rotational motors. With these actuators, the only consideration is output, since we've already established they provide no input. A motor's output is force. If you want to know how much force a given motor can exert, you need to find its power. When we talk about a motor's power, we mean the motor's rate of doing work. You do not want to confuse power with the energy supplied from the battery. The power of a rotational motor is calculated using both torque and speed, with a simple formula:

Power = Torque × Speed.

Torque is how much rotational force the motor can supply, also called turning force. Speed is how quickly the motor can turn. This is usually measured in rpm, or revolutions per minute. There are many other variables, like the efficiency of the motor and the input voltage, which can turn this simple formula into a maze of symbols. Since most of your applications will be with standardized systems like VEX, the simple formula is more than adequate for your uses.

The most important thing this formula tells us is that in most cases, speed and torque have an inverse relationship. Generally, the faster you make your motor rotate, the less torque it provides. Conversely, as a motor has to provide more torque, its rotation slows proportionally. Many actuators can provide variable torque or speed by changing the amount of energy supplied to the motor. The ones you will use in school very likely have that ability, so we'll cover them individually.

Electric Actuators

Electric actuators are the most commonly used actuators for school robots, but even within the category, you can still have many choices. With most systems like VEX or EV3, the type of motor has been chosen for you and is standardized. While standardization is very useful, especially for prototyping and designing your robot, a deeper understanding of how the actuators work is still important. In the first chapter, we showed you what these DC motors look like when you take them apart. Now it's time to take a look at them again, this time to see how you can build with them.

Rotational DC Motor

DC motors are one of the oldest types of electric motor. They are often one of the simplest motors to manufacture, so they are usually reasonably priced. Their simplicity also means they are one of the easiest motors to use. As most of our robots use batteries, the DC (direct current) motor is a standard. These motors convert electric energy into rotational motion. The speed of the rotation, or rpm, can vary greatly, depending on the voltage and the design of the motor itself. Every motor has a range of speeds but is actually only efficient over a smaller range, called the **power band**. We keep the motor's function within this power band, and control the speed within it, using a controller. Although these simple motors provide a range of speeds, they give no feedback regarding their speed, which must be measured in other ways, if required. Most rotational motors operate with high speeds but comparatively low torque, and they will only increase torque when bearing a load. If your motor is not powerful enough to provide the torque required, it is likely to **stall**. Stalling occurs when the load on the motor is greater than the possible torque. A motor running while stalled may overheat or in the worst case, burn out due to the high current running through the circuit. The motors in VEX and LEGO Mindstorms have a fuse to prevent permanent failures. To prevent or fix stalling, torque can be increased using gear configurations.

TIME-MANAGEMENT TIP
It is not in your best interest to allow the fuse in your motor to trigger, as resetting it is time consuming. It could be a make-or-break issue for a competitive team.

Small differences begin to come into play here. Brushed DC motors, typically two-wire motors, can spin in either direction and can be reversed by simply swapping the positive and negative wires. The simple DC motor discussed earlier is a brushed motor because it uses mechanical contacts, or brushes, to change the direction of the electromagnet's current as the motor spins. Depending on the way the motor is mounted on the robot, one spin direction will be regarded as forward, and the opposite as backward, but there is often no intrinsic reason for any particular direction to be favored. Brushless DC motors have three wires. The third wire is used for control because there is no physical contact dictating which direction the current is flowing. Instead, an inverter inside the motor itself changes the current to AC, or alternating current. The inverter is dependent on a small electronic controller, which dictates when the current will reverse. Slightly more expensive and delicate than brushed motors, brushless motors are also somewhat more powerful and less liable to wear. Both of these motors are comparatively quiet and fairly efficient for most purposes.

These motors are commonly seen in VEX systems. The one you will see the most often is the VEX 2-Wire Motor 393, discussed earlier. You will also hear this actuator referred to as a continuous motor, as it is capable of rotating continuously.

Servo Motors

Servo motors have less power than the continuous motor. They contain a rotational motor, but in addition, they have some type of rotation sensor. This provides accurate position feedback that can be sent to the program the robot is running. The sensors are called encoders. Many servos use an encoder that can measure how far the shaft of the motor has rotated; others use a sensor that shows the spot on the shaft to which the motor has rotated. Still others use encoder wheels, or gears, that help measure the rotation. These wheels have slits in them to let light shine through. A sensor detects the changes in light as the wheel encoder's slits are moved when it meshes with the pinion wheel. Counting the slits allows your program to calculate how far the axle has rotated. The **accuracy** of each of these motor encoders varies, as does perception of the accuracy. Some users claim that the LEGO Mindstorms EV3 motors are accurate to 3 degrees, though LEGO itself gives the accuracy as 1 degree. In practice, on a competition mat, performance might be very similar in either case. It is important to note that unlike VEX, LEGO does not have any dedicated servo motors. Every LEGO Mindstorms motor is a continuous motor equipped with some form of encoder. It is capable of giving feedback, but your program does not have to read it. VEX's servo is one of the few in school robotics that is not continuous. It can only rotate about 100 degrees, or by the program's definitions, from position −127 to position 127, as read by its encoder.

The motor itself does not know how far it has rotated, but the program can make use of the information sent to it to make accurate movements or measurements. For instance, the robot can be programmed to carry out an action until the sensor has read a certain distance or angle. For an example, see the program below.

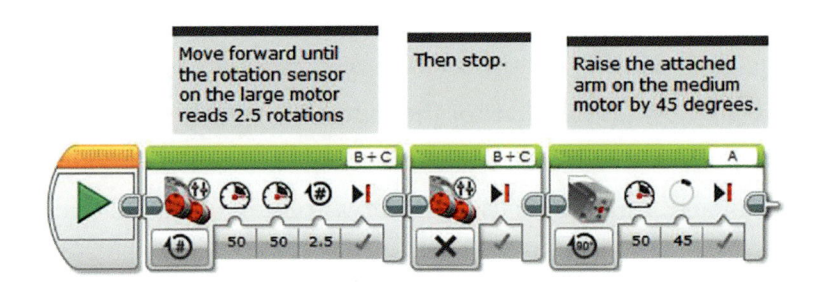

An EV3 program using the rotation sensors.

While it is a little strange to think of a sensor as part of an actuator, servos have incredible utility. They are often used for short-range, high-precision tasks, such as claw manipulation and arm controls. A bonus with VEX servos is that they can be set to hold a certain position, rather than having to continually adjust. In industry, some servos like to be calibrated and need to find their maximum positioning by going to that position each time they are powered on. With VEX and LEGO, this is not an issue. However, keep in mind that no two things are precisely identical, and take care when designing systems that need to be symmetrical. Check your robot carefully, and test your servos to be sure they're rotating the same amount.

Stepper Motor

Standard DC motors rotate continuously when the power is on. Stepper motors, as one might expect from their name, rotate one "step," or predefined increment, at a time. The electromagnets in these motors have teeth, like gears, and power is applied to each electromagnet at different times. With each pulse of energy, as one electromagnet is turned on, the gear moves one step to align with it; when that magnet is turned off and the next receives power, the gear moves another step to realign itself with the new magnetic field. This means that the motor moves step by very accurate step, with each complete rotation being divided into a specific integer number of steps of the same length. If the rotation is divided into, say, two hundred steps, then each step will turn 1.8 degrees (360 degrees/200 steps). Stepper motors can thus be used for very precise positioning or speed, even though there is no encoder or feedback device in these motors.

KEEP THE POWER ON

When you use a stepper motor, remember that the only thing maintaining its position is an electromagnet, which cannot hold a position when unpowered. You have to write a program so the power stays on until a task is complete.

The motors are generally reliable, and their accuracy in positioning means that they are used where precise positioning is very important, such as in 3-D printers. Another advantage they have, which regular DC motors lack, is that they generate high torque at low speeds. Furthermore, since the motor can be stopped on a specified step, many models can even maintain torque when they are not moving, which means that robot arms, for example, will be held in place even when not in motion.

Linear Actuators

Electric Linear Actuators

Electric **linear actuators** are fairly uncommon, on school robots and elsewhere. Most "true" electric linear actuators work on a complex system of electromagnets, on the same principles as a brushless DC motor, but they are much more delicate. In order to create the linear motion, a much more involved controller is needed. They're not yet widespread, or powerful. Instead, a linear actuator is most often a DC motor, with a gear train that is converting rotational force into linear force mechanically. These are very commonly used in industry and elsewhere, and on a number of school robots. Mechanically converted linear actuators can be of several designs. One of the most prevalent uses a worm drive, which contains a moving shaft within an outer shell. They work in a similar way to a servo motor, as they contain a DC motor inside the outer shell, with a potentiometer for feedback. The shaft is threaded like a screw. In fact, it is a perfect example of the simple machine.

The shaft is moved in and out of the shell by a **worm gear** attached to the motor, shown below:

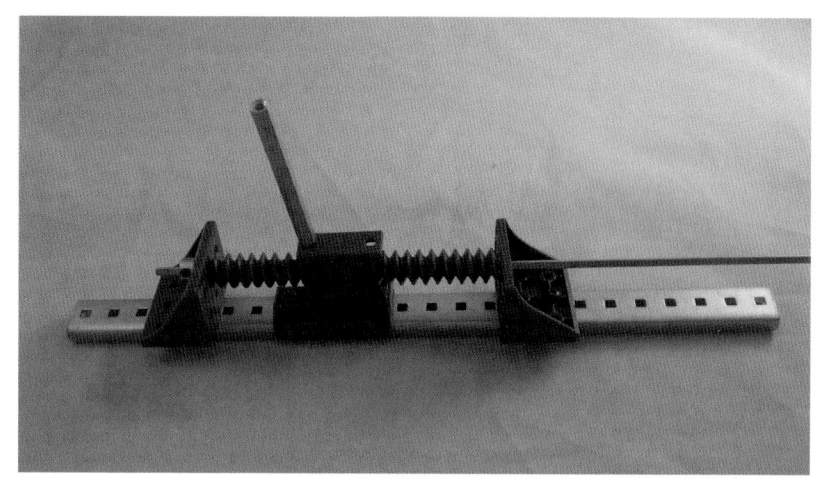

A spindle of VEX worm gears creates linear motion.

These work slowly but are very accurate, and they can lift heavy loads. A big advantage is that they have a lot of holding torque when they are still, and they maintain that holding torque even when there is no power. These linear actuators are sometimes used to build simple robot arms for heavy lifting. This type of actuator is very common in industry and manufacturing. Some companies have even created versions that are based on stepper motors and can provide linear motion in the same precise pulses as that actuator.

There are several other methods of converting rotational force into linear force. A **rack and pinion** system, one of the other most commonly used, is similar to the worm drive but has several important distinctions. The rack is a long, flat bar with teeth on it. The teeth on the rack mesh with the pinion gear on the motor and push or pull it, usually along a channel or guide. Generally, the pinion is stationary, propelling the rack. A particularly useful application is for lifting. A rack attached to something like a forklift or gate can give you extraordinary power and control over your lifting mechanism. It is not unusual for rack and pinion systems to be used to propel a system across the rack. There are fewer applications for this in school robotics; once again, industry has a greater range. Rack and pinion drives have gained traction with 3-D printers and similar linear, highly calibrated systems. Both the worm gear and the rack and pinion gears are available for VEX and LEGO Mindstorms.

If you've been paying attention, you might have noticed a small problem. We have defined the power generated in terms of torque, which is rotational force. By definition, a linear actuator won't have rotational force, so how do we calculate the output of a linear actuator? You've already seen the relationships of power, torque, and speed. But how does that translate to linear motion? The answer is that it doesn't! Or rather, it doesn't need to. The two types of linear motion we've already covered use one very important thing: a wheel and axle, or, again in our case, a pinion gear on a shaft. Since the initial output is rotational force, we can simply use the formula for torque. We established earlier that torque was turning force. However, if you look at the formula for torque, you'll find the part that wasn't mentioned before—distance:

$$\text{Torque} = \text{Force} \times \text{Distance}.$$

Torque is measured as the amount of force applied at a distance from the point of rotation.

We are trying to find the force, so we need to rewrite the formula as:

Force = Torque/Distance.

The pinion gear on a shaft is essentially a **lever**. However, since we're talking about torque coming from the point of rotation, we need to look at it as a lever in reverse. Force is the missing variable; torque is the measured output of our motor. Distance is always the **radius** of the gear; that is what is working as our lever.

Distance in this case is distance from the point of rotation—that is, the radius of the pinion gear. Before substituting any numbers for the variables, we must consider units. The formula, once again, gives it away. Torque is measured in force times the mechanical advantage of the distance. That means that torque should have units of foot-pounds, ounce-inches, or if you're working in SI, newton meters. Be sure to be aware of your units: if torque is in foot-pounds and distance is in inches, you'll have to convert one or the other, or your numbers just won't match up. Of course, pounds, ounces, and newtons are all the units of force for the equation. If you keep all this in mind, you will be able to calculate the force you'll achieve with your rack and pinion setup, whether you're trying to lift a weight or push an object.

Cam and Follower

The last, possibly least common type of mechanical linear actuator is the **cam** and follower. The cam takes the place of a pinion wheel, and has a smooth surface. The most common type is a disk cam. Rather than being a perfect circle, the cam is irregularly shaped, often presenting a profile like a pear or a snail shell. The follower is a shaft, held perpendicular against the cam by either gravity or a spring, and passing through a support with a bushing. Bushings are a type of bearing, specifically for where an axle or shaft passes through a support. Bearings are any surface or mechanism designed to resist **friction** and abrasion. As the motor rotates, the follower is pushed out through the bushing by the eccentric radius of the cam and then falls back in. While this is not as precise as other types of linear actuation, it can happen much more quickly and repeatedly. Cam and follower systems are useful for any sort of tapping motion, as a release mechanism or a **switch**, and of course, for whatever new and exciting things you can dream up. This configuration can also be used in combination with other linkages, especially to create holding systems without a slower, dedicated grabber. The photo on the opposite page shows how a cam can move a follower shaft up and down.

    Mechanical linkages are a great way to transform motion. All of the options we've reviewed above are possible with both VEX and LEGO Mindstorms. As long as you take care to design within the parameters of the kit you are using, there are always fun and interesting ways to accomplish your tasks. Be observant, and innovate.

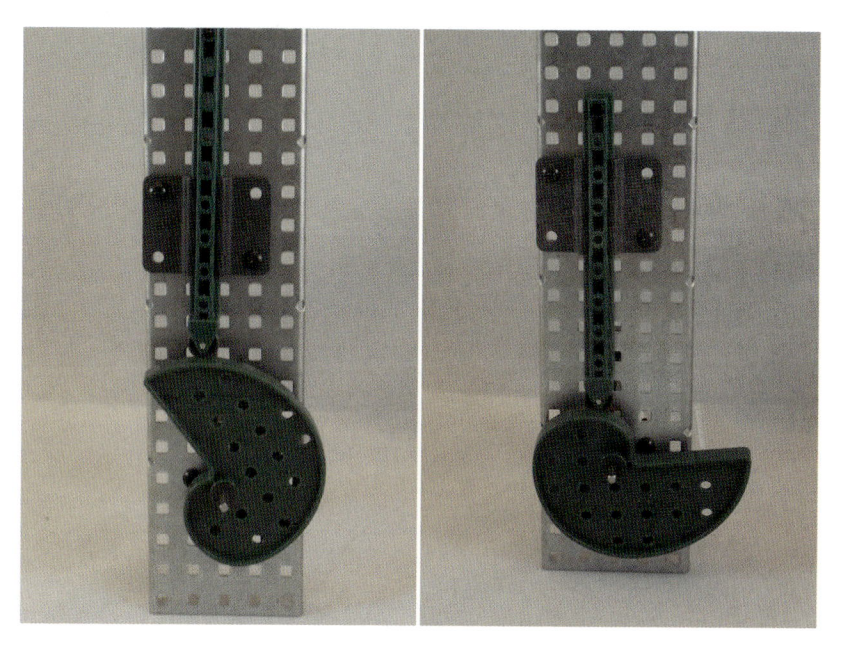

The rotation of the cam moves the follower up and down.

Pneumatics and Hydraulics

Pneumatics and **hydraulics** are often confused with each other. Hydraulics refers to the movement of liquid through a system; pneumatics is anything that uses air pressure to do work. When we are talking about pneumatics and hydraulics for robots, we are specifically looking at actuators and their support system. Hydraulic actuators are almost always in a single form: the hydraulic cylinder. They are widely used for heavy-duty, single motion mechanisms. They actuate by forcing pressurized liquid against a piston within the cylinder, moving the piston out to create linear motion. Systems like this are called single acting. If you added another port to add pressurized liquid on the other side of the cylinder and force the piston back down, you

LEGO EV3 GRABBER

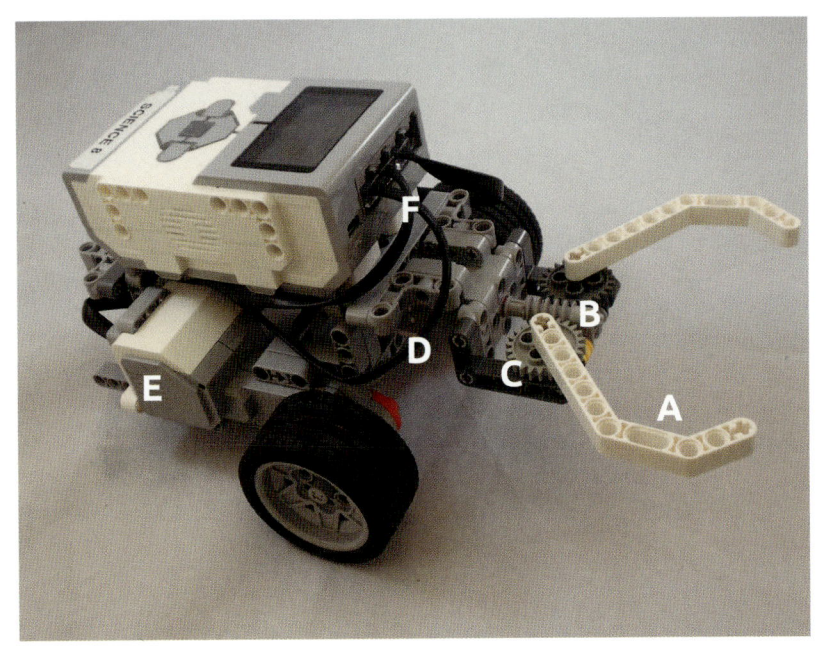

An EV3 robot with a two-arm grabber moved by a worm gear, as designed by Dr. Damien Kee.

This is a simple EV3 grabber that is simple to construct. It uses one motor for movement and another to control the grabber. It can easily be adapted for other uses.

A. A simple grabbing arm, made of two angle beams. These could grab larger items, but could also secure and retrieve loops.

B. A worm gear, slid onto an axle that leads into the medium motor. The worm gear turns as the motor turns, but changes the rotary movement of the motor into the horizontal direction of movement of the arms. The worm gear also provides increased torque.

C. One of the two spur gears that mesh with the worm gear and help to change the direction of the force. The spur gears control the movement of the arms.

D. The EV3 medium motor is hiding behind all the angle beams and beams with pegs. Because the axle only goes a very short distance into the medium motor, a sturdy structure is required for attachments using this motor.

E. The EV3 large motor, powering a wheel.

F. Cables connect the motors to the output ports so that the robot brain can send instructions out to the motors.

would have a double-acting hydraulic cylinder. The pressure system also means they are stunningly powerful; they are found almost universally in heavy machinery. You might even find one actuating your car's emergency brake. However, in the scale of LEGO Mindstorms and VEX robots, the scale of equipment and power necessary for hydraulics is simply nonsensical. Fortunately, we have a better option. Pneumatic actuators are considerably safer and less demanding than hydraulic systems, and they have plenty of power for these robots' needs. We all know that gases are less dense than liquids. Since there's more space between molecules in a gas, it takes less energy to fit them in. Have you ever tried to get on a crowded elevator? It's very similar: it takes more work to get into a space, not because others are occupying the same space, but because they're so close to it.

Pneumatic systems, being by some standards less demanding than hydraulic systems, come in many more varieties. One of the most fascinating of these is a flexible pneumatic system that mimics the expansion and contraction of muscle. For the rest of us, pneumatic cylinders will do the jobs we need. These cylinders can also come in single-acting and double-acting varieties, and they function in a similar way. The thing is, that function is radically different from what we've discussed before. The biggest mental hindrance for most teams is that naturally you don't power pneumatics with a motor. So where does the energy come from, and how do you control it? Pneumatics is powered by a compressed gas. Commercial setups often use gases like nitrogen. In the EV3 and VEX environments, it's regular old air, pumped into a storage tank or reservoir, just as you would pump it into a tire.

SAFETY TIP

It is important to observe all safety standards; high pressure anything, even air, can be dangerous. When working with pneumatics (or filing metal, or anytime you are building), wearing safety glasses is an excellent idea.

VEX systems also have a shutoff switch that can be used to work on a system without losing total pressure. This should be fully utilized, and carefully: the switch must be installed a certain way. In the absence of that, or when the shutoff switch is open, pressure is maintained in the reservoir by means of a valve. In EV3, the valve is controlled by a mechanical switch. VEX takes things a step further and uses a **solenoid**. Solenoids, more properly known as **solenoid valves**, are actuators in their own right: they, too, create motion from energy. A solenoid valve sits between the pneumatic cylinder and the tank, with its valve tightly closed. When we want the cylinder to actuate, we send power to the solenoid valve.

Now we can see where the solenoid valve gets its name: it contains an electromagnet called a solenoid, which lifts the valve and allows gas through. Single-acting cylinders usually use a spring or other mechanical system to return to their original position. As long as there is adequate pressure in the tank, the actuator can be triggered again. When using a

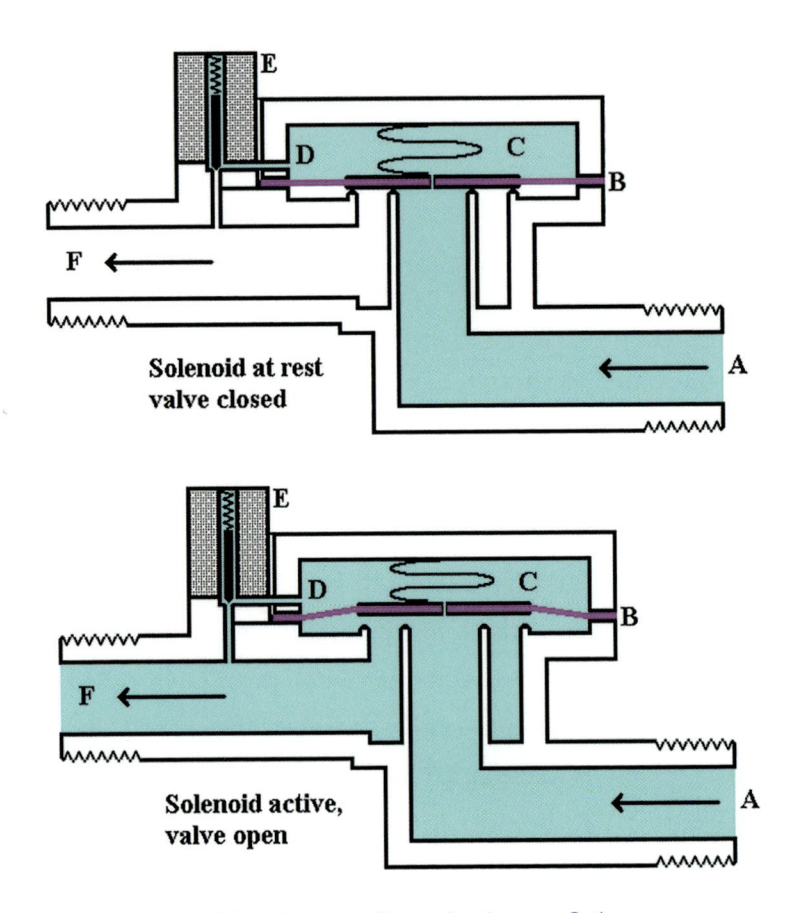

A solenoid valve, with A, the input; B, the diaphragm; C, the pressure chamber; D, the pressure relief passage; E, the solenoid; and F, the output.

double-acting cylinder, a second solenoid must be triggered to send gas to fill the opposite chamber in order to retract the arm. Each movement of the arm is called a stroke. Since they rely on air pressure, the only way to make a half-stroke is to reduce the pressure using the pressure regulator. You cannot make a half-stroke, and then a full stroke. The only two positions are extended and retracted. Given their all-or-nothing states, pneumatics are well suited to quick lifting, throwing or striking, and pushing.

The LEGO pneumatics add-ons are not very well known, but they can be useful as they add an extra power source if you find yourself limited by the number of motors you can use. They are also very powerful, so they can accomplish lifting tasks that the regular motors have problems with.

The parts available include tanks for compressed air, hoses, switches, T-joints, pressure gauges, and actuators. LEGO pumps are used to fill the tanks with compressed air, which is closed using a switch. When this switch is opened, air is released into the hoses and moves to the actuator, causing it to open or close.

LEGO makes two sizes of pump—a large, spring-loaded manual pump that is easy to use, and a smaller pump that requires more effort but that can be connected to one of the

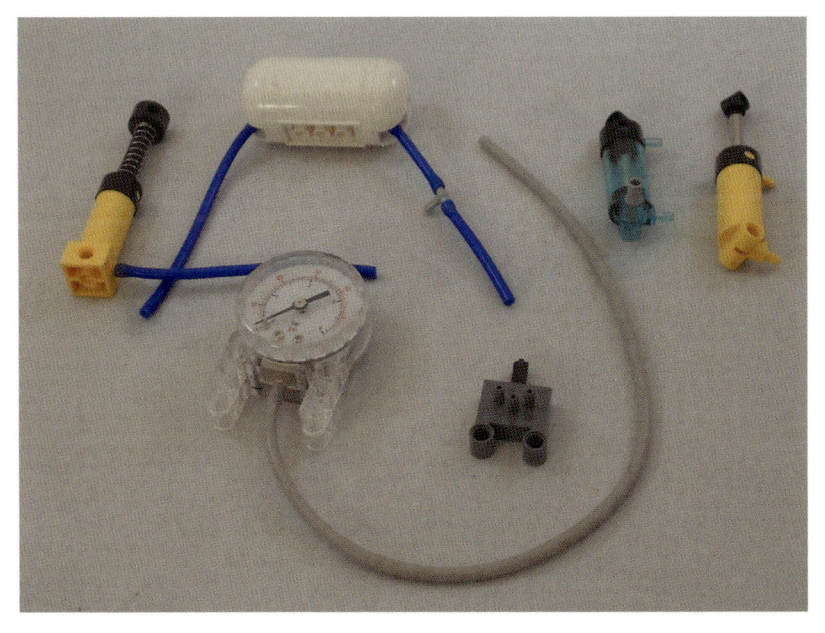

LEGO pneumatic pieces (clockwise from left): pneumatic cylinder; tank; pump, another cylinder; valve/switch; and a pressure gauge.

Right Tool for the Job 33

robot's motors, which pumps it automatically. The pneumatic switches have three ports for air and three positions for the switch. The middle position is the off position; it prevents air from flowing through any of the ports. When the switch is pressed down, the middle and lower ports are opened to allow air to flow through. When the switch is up, the middle and upper ports are opened to allow air to flow through.

The pneumatic pieces are connected by air hoses and T-joints, which can extend or split a hose. The tanks can be filled with air by hand before the robot's programs are started, and the switch is then activated by the program. This gives you the advantage of freeing up a motor port for other functions. Alternatively, the smaller pump can be used to fill the tank mechanically during the program using one of the motors.

The actuators themselves are linear actuators, with a shaft inside an outer cylindrical casing. The compressed air is used to move the shaft in or out. Some actuators have only one air input, which causes the inner shaft to extend when air flows into the actuator, so these would have to have the shaft depressed by hand or using a spring. Other actuators have one input at the top of the casing and one at the base. Air going into the top input will make the shaft retract, while air flowing into the input at the base of the cylinder will cause the shaft to extend.

ALL WORK, NO PLAY

Your motor must be mounted securely with a minimal amount of play, or movement. The more play or wiggle room you leave, the easier it is for motors to wear out, fall out, or get caught and stall.

The pneumatic switch is quite easy to trigger, so the robot can be programmed to move close to an obstacle to trigger the switch, allowing air to flow and extending the linear actuator, allowing it to move or trigger an attachment.

Independently, each of these systems is nothing more than a nifty gadget. Actuators can perform a wide range of actions, but it's important to be able to choose the right one for the task. The actuator must be able to accomplish its task: turning energy into motion. But to be functional in a robot, actuators must be able to complete the same action reliably, over and over again.

These are only concerns for quality control. The real challenge with actuators is right in the book's title—they must provide the power to do tasks. The next chapter deals with two things: how to accomplish tasks with the least amount of power, and how to provide that power in such structured systems.

This vehicle uses a low center of gravity and treads that generate friction to resist being pulled backward.

3 Accomplishing Your Tasks

There's more than one way to skin a robot. Too often, builders tend to overlook a major point in the creation of their robot. Most people will focus on gearing motors down and creating the torque necessary to perform the tasks they've decided on. Rather than jumping in, it is worth taking the time to see if there is a way to make the process more efficient; that is, to reduce the power necessary to accomplish a task. Power is impressive, but finesse can be truly awe inspiring.

There are several considerations to managing the power needs of your robot. The quick and dirty list runs like this: **traction**, friction, and mechanical advantage.

Traction is a friend of friction. Friction is the tendency of surfaces in contact with each other to resist movement. Think of dragging your shoes on concrete. It's not as easy as walking while picking your feet up, and it sure does slow you down a lot. Friction is everywhere in your robot: at attachment points when they get jostled, between gears, and between the tires and the ground. The bushings and bearings we put at moving points have just one job: to bear friction. They do that either by being a smoother surface, which cuts down on friction, or simply by protecting other parts from the abrasion that goes along with constant friction. The more you can reduce friction—whether it's by greasing the gears inside your motors, adding bushings, or choosing the straightest axles—the less power you waste compensating for it. A lighter load on an actuator will also decrease friction. Motors always work against friction; avoid it as much as possible.

With one exception, that is.

If friction is like dragging your shoes on concrete, traction is like standing on a slide. If your shoes have enough traction, gravity cannot pull you down the slide. A robot's wheels maintain traction the same way your shoes do—by being grippy, or having friction with the surface they sit on. The more traction your robot's wheels have, the less likely they are to slip. That means your robot can go farther in a single rotation of the actuator, and that is definitely efficient. If you can't have wheels with more grip—for example, if you are using omnidirectional wheels—then you've got one other option. Just as a lighter load on an actuator will decrease friction, a greater load will increase it. When you're trying to increase traction, a heavier bot is a good thing. The more force that is acting on the wheel—typically the robot's weight, or gravity acting on its mass—the greater the friction it has with the ground.

Friction is everywhere moving parts intersect. Traction is only where parts intersect with the ground. Mechanical advantage exists anywhere you use a simple machine to make a job easier.

WEDGE
(one of the six simple machines)

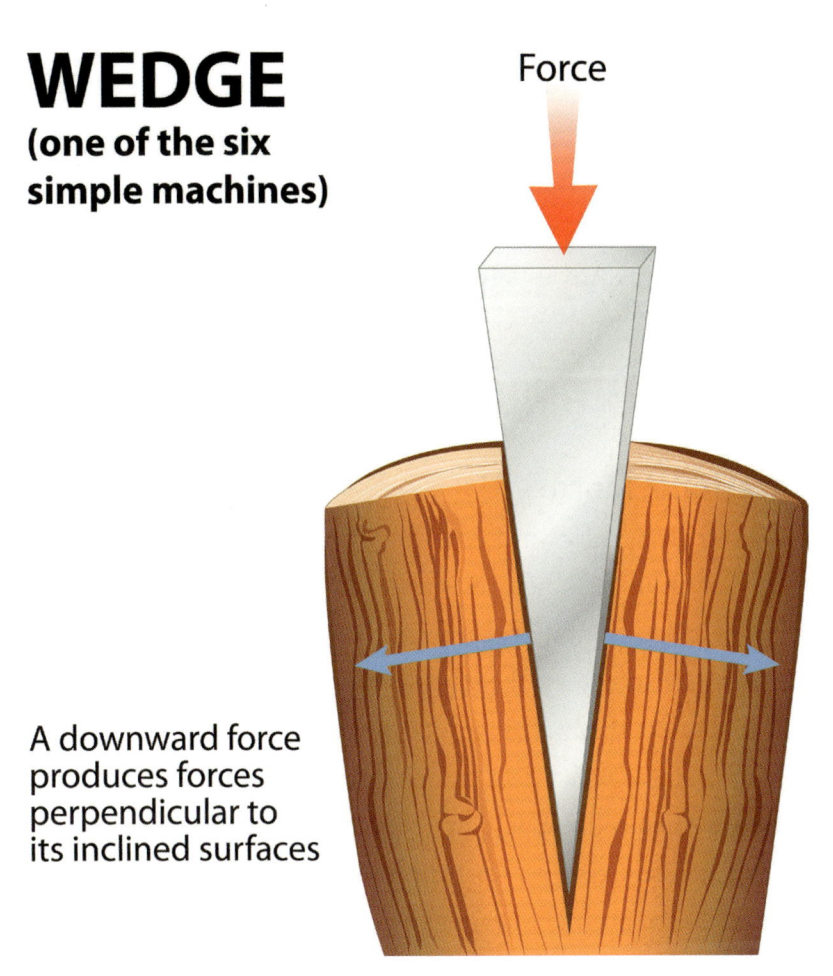

Force

A downward force produces forces perpendicular to its inclined surfaces

This wedge provides a mechanical advantage for splitting this log. In robotics, wedges can make lifting objects easier.

Because the mechanical advantage of a wedge is the length of the sloped side divided by the width of the wedge, if I hit a 2-inch-long (5-centimeter) wedge with 1 pound (0.45 kilograms) of force on its 1-square-inch (6.5-square-centimeter) blunt end, I get approximately 2 pounds (0.9 kg) of force. Being aware of simple machines and the advantage they can provide is the first step to a very efficient robot.

Every wheel and axle, screw, and lever on your robot should create some form of mechanical advantage. Mechanical advantage is simply a ratio of how much work you put in to how much work you get out:

Mechanical Advantage = Output Force/Input Force.

Moving Around

Rotational Motors

The most straightforward use of rotational motors is to move the robot around. When you create a system that allows the robot to move around, it is referred to as a drivetrain. This includes the actuators, the wheels, and whatever means you use to connect them. This can be anything from a simple axle driving a single wheel to a complex series of gears. You can even use a chain drive, which operates like the chain on your bicycle, transmitting power from the pedals to the wheels. One of the greatest challenges in designing your drivetrain is the abundance of options. You need to decide how you want your robot to move. Will it be on omnidirectional wheels? With a crab drive? Or will you just skid steer? The first step is to determine your needs. If you need maneuverability, omnidirectional wheels allow you to zip about the field, at the expense of traction. For a drivetrain that needs to support and transport a heavier robot, larger wheels with greater traction may be preferable. Most often, these styles of drives use a power train that skid steers.

Designing a power train for a LEGO motor is relatively simple. Many LEGO robots use a **differential steering** system, where each motor controls one wheel of the robot, so that the wheels operate independently and can move at different speeds. Drives of this type are also called skid steering, both in robotics and in industrial applications for heavy machinery. Skid steer systems are prevalent in VEX just as they are in LEGO, and are conceptually interchangeable.

If you imagine steering a wheelchair, you will get the idea. If both wheels move in the same direction at the same speed, the robot will go forward. However, if one wheel moves more slowly than the other, the robot will start to turn in the direction of the slower wheel because the faster wheel is moving a greater distance in the same time. The greater the difference in the speed of the wheels, the tighter the turn will be. This type of steering allows a robot to make steering changes quickly and efficiently.

This is different from the way a car is driven. Cars use a **differential gear** or steering drive system, where two wheels that use the same power source are able to turn at different rates. You can build robots with a differential gear system—LEGO has a very good differential available—but it is more complicated than the differential steering. The advantage of a differential gear system is that because the motor has been separated from the steering, you do not have to worry about matching the motors. On the other hand, you can no longer turn on the spot, as you can with the (wheelchair) differential steering system, which is the most common in LEGO robots.

Moving Forward and Backward

Let's think about a LEGO robot moving forward using a differential steering system with two motors, each controlling one wheel. Motors tend to run slightly differently from each other, so often robots which you would like to run in a straight line instead curve off course slightly. If you have extra motors, you can minimize this annoying habit by matching your motors.

Matching motors is often necessary only for competitions, where there are very narrow margins for success. Usually a little drift is not serious, and the worst of it can be controlled by using the Move Steering blocks. These synchronize your motors to a great extent.

SYNCING THE WHEELS

Attach two motors to your LEGO brick using long wires. Connect the two motors using one long axle. (This is usually a poor idea, but in this case, it will help you see which motors work at the same **rotational speed**.) Lay the motors flat, program them to move forward using the Motor (NXT) or Large Motor (EV3) blocks, and run the program. If the motors run at different speeds, one of them will jump and the two motors will no longer lie flat. If this happens, remove one of the motors, and try another, until you find two that match, lying flat at all speeds. You can then test for rotational accuracy by attaching a gear, with a peg inserted in one of the gear holes, to one of the motors. Run the motors for a specified number of rotations (try a few different numbers), and make sure that the peg always ends up in the same position from which it started. Remember not to use Move blocks (NXT) or Move Tank/Move Steering blocks (EV3), as these are designed to synchronize the motors and will mask problems.

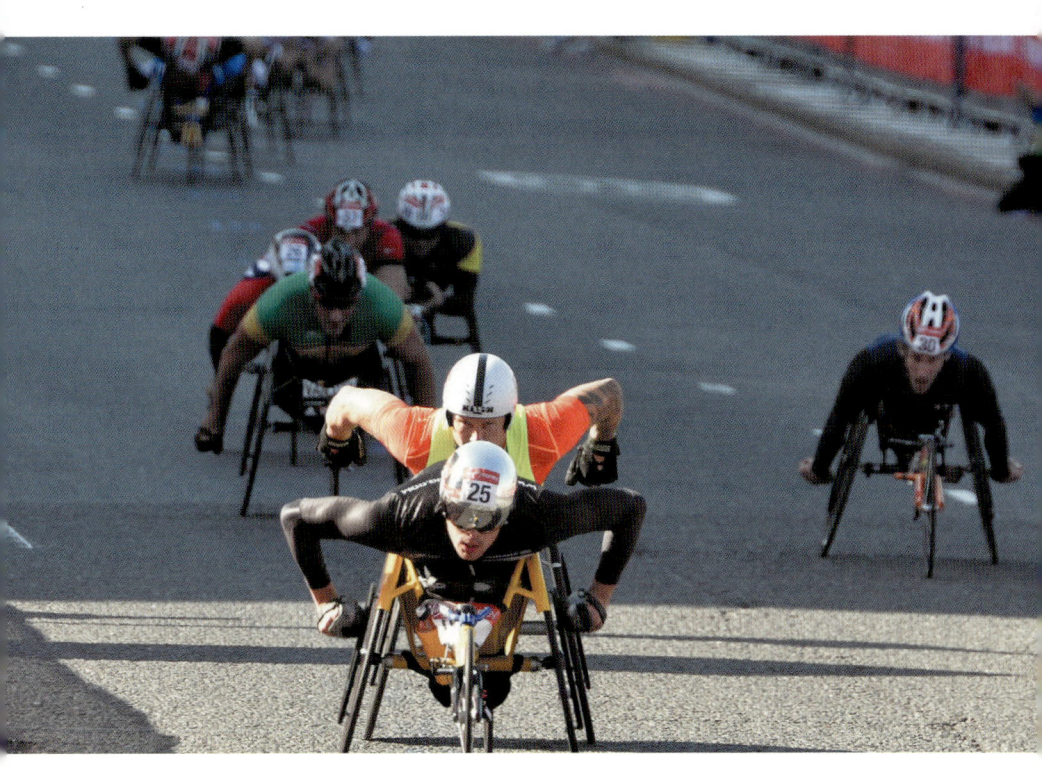

Wheelchair racers spin their wheels at different speeds when they have to navigate around a turn.

Once we have a robot with a differential drive, moving forward is easy. Simply program the robot to move with both motors running at the same speed.

Moving forward for the VEX Motor is just as easy. Give the program a power value, represented as a number in the integer range –127 to 127, with negative integers being reverse and positive integers driving forward. It's worth noting that the range is not random. In computing, $2^7 - 1$, or 127, is the highest value for a signed eight-bit integer. "Signed" means it can be positive or negative. That's why both VEX motors use these values: the 393 as a measure of power, and the VEX servo as an absolute position.

Accomplishing Your Tasks 43

ROBOTC, the programming language used for the VEX robots, does not have a command that synchronizes the two motors. If your robot is swerving a little instead of going straight, another possibility is to reduce the power on the faster-moving motor slightly until both motors are turning at the same speed, though different powers are being used. So if your robot is turning to the left, which means that your right motor is going slightly faster, reduce the power level of the right motor bit by bit, until you find the level at which the robot moves straight.

Speed is a relative term when using these robots. The speeds given in the LEGO Mindstorms programming GUI are percentages of the greatest possible speed, ranging between 0 (stopped) and 100 (the fastest speed). Be aware that the absolute speed—how fast the robot will go in centimeters per second—depends on several variables, including the size of the wheel and the charge in the battery.

Let's look at these two variables. The charge on the battery is self-explanatory. You can see intuitively that as the battery runs down, the motor will move more slowly, in the same way that a flashlight bulb dims as the battery runs down. Slower rotations result in slower speed, even though your program still specifies a speed of, for example, 50.

In contrast, you can control the size of the wheels you use, and in that way, control the effect on your speed. You need to understand the relationship between the **diameter** of a wheel and its **circumference**. The circumference is the distance around the edge of a circle, or in this case, the wheel. So, as a wheel rolls through one rotation (or rotates 360 degrees), it

MATH MOMENT

To measure the distance around a wheel, you use the formula for circumference:

$$\text{circumference} = 2\pi r$$

where r is the radius of the wheel and π is a constant approximated by 3.14.

The radius is the distance from the center of the wheel to the edge of the wheel.

Because 2 × radius equals the diameter of a circle, this formula can also be written as:

$$\text{circumference} = \pi d$$

where d is the diameter of the wheel.

The diameter is equal to the distance across the wheel through the center, or the widest distance across the wheel.

If you need to make your robot travel a certain distance, let's say 35 centimeters (14 inches), you can calculate how far the robot will travel in one rotation, and then work out how many rotations the wheels need to make to go that distance. The standard wheel for the EV3 robot has a diameter of 5.6 centimeters (2.2 inches). If the wheel diameter is 5.6 centimeters, the circumference of the wheel (circumference = πd) is 3.14 × 5.6, or 17.58 centimeters (6.9 inches).

We know now the robot will move about 17.5 centimeters when you program it to move one rotation (or 360 degrees). If we want the robot to move 35 centimeters, we find the number of rotations by dividing distance by circumference. When you divide 35 by 17.5, you get two rotations. Computing rotations makes your programming much easier, as you will not waste time guessing to see if your robot is traveling the right distance.

travels through one circumference. As a wheel rolls through one rotation, it must travel the same distance as the distance around its edge. It follows that a robot with big wheels will travel farther in one rotation, or in one second, than a robot with smaller wheels—that is, it will go faster.

So, if you know the circumference of the wheel, you can calculate how far the robot will travel when you program it to move one rotation, or 360 degrees.

We have discussed moving, but we also need to consider stopping. When you program the robot to move for a certain number of degrees, rotations, or seconds, the motors will stop after that number of degrees, rotations, or seconds is complete. This does not mean that the robot will stop. Imagine that you are pushing a toy car. When you stop pushing, the car continues to move for a while, as it still has some momentum. If you would like your robot to coast a little after stopping, you can set the Move Steering or Move Tank blocks to "coast" in the last section of the block.

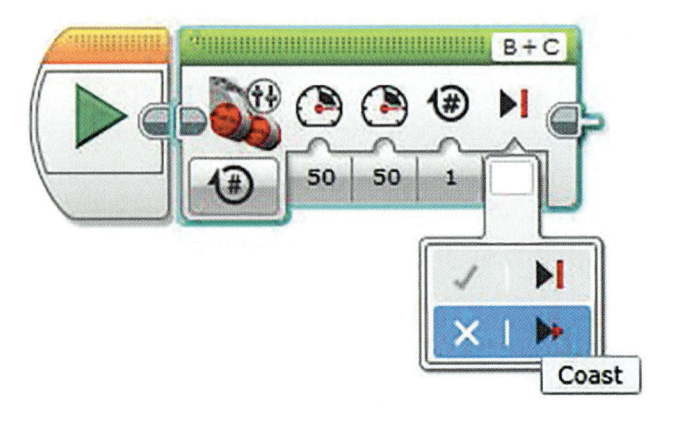

This EV3 program directs the robot to move forward and coast at the end.

More often, if you need it to stop exactly when the movement is complete, you should set the Move Steering or Tank block to "brake," and you could even add a Move block set to "off" to make doubly sure.

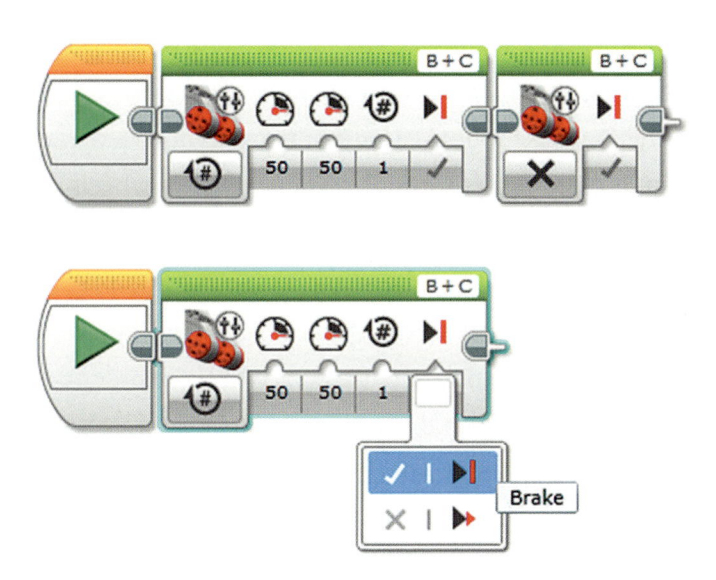

These EV3 programs direct the robot to move forward and brake. The upper program has a block to stop the motors at the end.

Another actuator configuration is more a curiosity than a useful way of building a school robot—a **synchro drive**. The wheels are mounted on turntables. One motor controls the rotation of the wheels (tells them to turn, resulting in forward or backward movement of the robot), and a second motor controls the direction in which the wheels are facing, simultaneously. If the robot is moving forward, and it wishes to turn right, instead of the body turning right, the wheels will rotate through 90 degrees, and the robot will then move in the desired direction, but still moving forward, without the orientation of the body having changed. Only the orientation of the wheels changes.

For VEX, there is a little more room to explore. Teams that opt for omnidrives use a square chassis, with an **omniwheel** on each side. By pairing parallel motors to drive in synchrony and utilizing the mini wheels that sit at 90 degrees from the omniwheels, these robots can drive not only forward and back, but also left and right, without raising a pair of wheels or reorienting the robot. This can be a simple novelty or a huge advantage, depending on the year's game. Always play to your strengths, but also play to the game.

Drives that raise opposite pairs of wheels in order to switch direction are called crab drives, due to the way they scuttle side to side when their main drive is lifted. Of the many drive

How a robot would do a point or spin turn around the orange cross, which would stay in the same spot.

options, we're going to stick with the differential steering system for now.

Turning

One of the advantages of a differential steering system is that the robot is able to turn on the spot. A **point turn** (also called a spin turn or a zero-radius turn) is the result when one wheel rotates forward and the other rotates backward, at the same speed, causing the robot to sit and spin in place.

In the picture on page 48, both wheels move, and the orange cross stays in the same place.

This turn takes up the least amount of space possible. For an illustration of this, visit the author's YouTube channel at this URL: www.youtube.com/watch?v=TjAqyHIKbxM. The EV3 program for this turn is shown below.

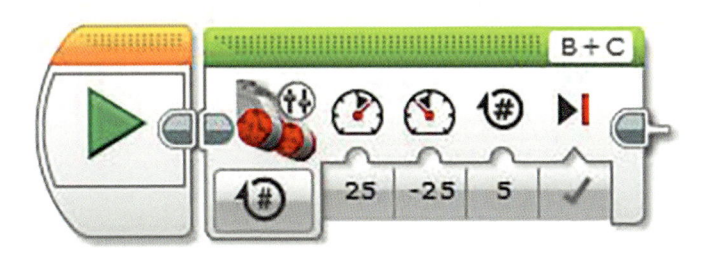

To program a spin or point turn, one wheel should move forward, and the other should move backward, at the same speed.

An EV3 program for a spin or point turn.

Accomplishing Your Tasks 49

In contrast, a **pivot turn** (or swing turn) results when one wheel rotates and the other stays in place, causing the robot's body to "swing" around the stationary wheel, taking up more space. To see an illustration visit the author's YouTube channel at this URL: www.youtube.com/watch?v=BCj8oRjt8Lc. With these turns, the drag or friction on the stationary wheel is called **turning scrub**.

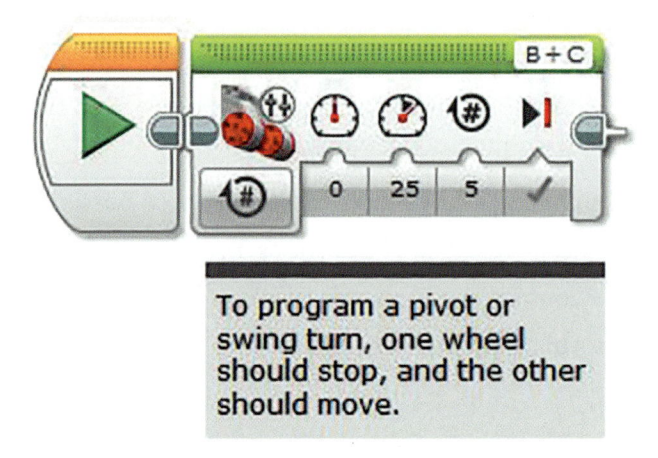

To program a pivot or swing turn, one wheel should stop, and the other should move.

An EV3 program for a pivot or swing turn.

Finally, the last type of turn the robot can make with the differential steering system is a **curve turn**, where both wheels move in the same direction, but at different speeds. The faster wheel will travel a slightly longer distance, so the robot will travel in a curve with the faster wheel on the outside of the curve.

50 Actuators and the Power to Do Tasks

> To program a curve turn, both wheels should move in the same direction, but at different speeds.

An EV3 program for a curve turn.

Lifting Motors

Robots are often required to lift things. This can be done in a variety of ways, depending on the object that must be lifted and the height to which it needs to be raised.

Rotational Actuators

Simple lever attachments on a rotational motor are often used to lift items, usually those that are not too heavy and that do not need to be lifted too high. They can be as simple as a beam attached to the motor. You would use this, for example, if your LEGO robot needs to lift a mission piece using a loop. Attachments this simple are useful even in VEX and can be used for jousting game objects off raised platforms. More intricate designs such as articulated lift attachments (two sections connected by a hinge or a joint) are also possible.

MATH MOMENT: HOW MUCH DOES MY ROBOT ROTATE?

It's a great temptation, when programming movement using rotations or degrees, to imagine that wheels rotating one rotation, or 360 degrees, will make the robot itself turn through 360 degrees.

You will find that it doesn't. Think about it: What is actually turning through those degrees? The wheel! And when the wheel turns once, the robot actually turns much less. Try it and see. So how can we calculate the number of rotations or degrees that makes the robot spin once?

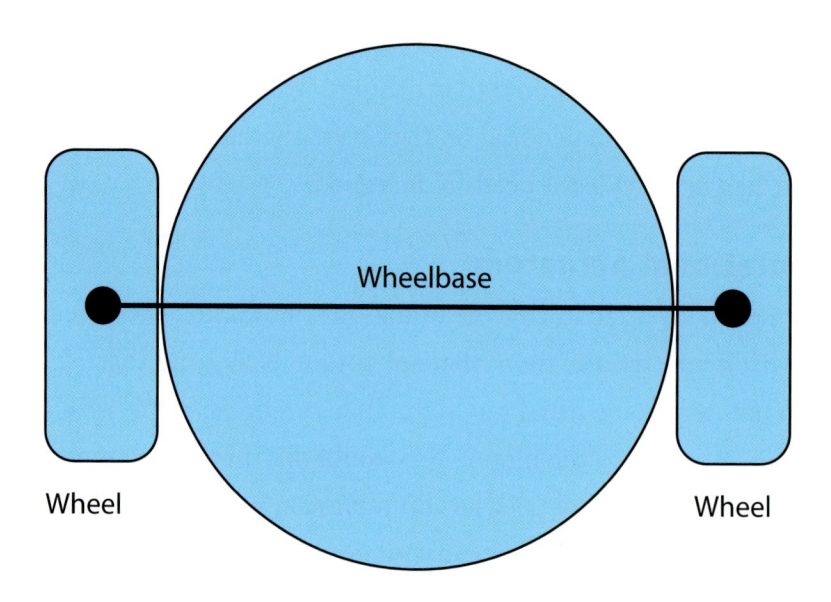

A diagram of a two-wheeled robot showing that the wheelbase is the distance between the center of its wheels.

Look at the diagram on page 52, representing a robot. The rectangles on either side are the wheels. The distance between the wheels is called the **wheelbase**. If the robot makes one full turn, any point on the outside of the robot will move a distance equivalent to the circumference. Looking at the diagram, we can see that the wheelbase can be regarded as the diameter of the robot. So, remembering the formula for circumference, we can deduce that the distance any point on the edge of the robot will move in one full turn must be equal to π × wheelbase.

The ratio between this distance and the circumference of the wheels gives the number of rotations the wheels must turn in order for the robot to rotate once.

$$\text{Rotations} = \frac{\pi \times \text{wheelbase}}{\pi \times \text{wheel diameter}}$$

$$= \frac{\text{wheelbase}}{\text{wheel diameter}}$$

So, if you would like the robot to turn through 90 degrees, you would find out how many rotations it takes to get the robot to turn once (formula above) and multiply the number by 90/360 or ¼. A 90-degree turn is one-quarter of the distance around the circumference.

HANDICAP ACCESSIBLE

Let's say you've built your awesome robot, and boy, can it move. On a flat surface. Or the right flat surface. Have you thought about where it can move? Does it move through soft carpet as well as it does on tile? Do the tires have enough traction to go up an incline? Do the actuators have enough torque to go up a steep incline? Did you consider whether you have the clearance to go over a bump, how to keep your robot from losing grip, or making sure the gear train is out of the way, so it doesn't get caught on anything? There are a

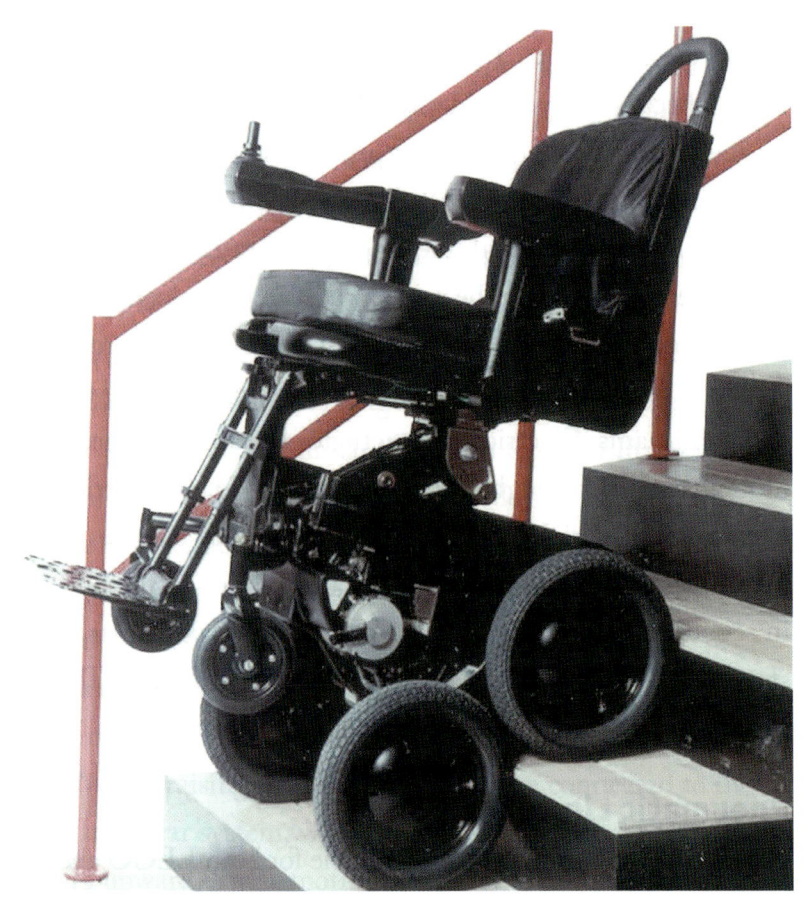

Dean Kaman's iBot wheelchair was able to climb stairs and to raise people so they can look others in the eye. A new version is being developed.

hundred more considerations. At least, that's how it is when you're navigating the real world.

Imagine having considerations like this in your everyday life, anytime you left your house. Is that doorway wide enough to go through? Are there elevators? There are a hundred barriers every time someone in a wheelchair leaves the house. So if we have the technology to build robots that navigate obstacles like these in schools, why aren't there more in the real world? Thankfully, the tides are turning. Dean Kamen is the creator of the Segway Personal Transporter. A man who reportedly has a collection of vintage wheelchairs in his house is exactly the person you might expect to revolutionize the wheelchair. He actually did it in 1999!

Kamen was not the first to ask, but he was one of the first to provide an answer to a question we're still struggling with today. Why design infrastructure that is accessible, but not accessible to everyone, when we can create technology to make the whole world accessible to anyone? His answer was a revolutionary design. The iBot wheelchair was revealed in 1999 and became available in 2003. It uses two small wheels on each side, connected to a bar, which itself can rotate. The wheels' ability to move end-over-end allows the person using it to climb stairs. While on a level, the wheelchair can use a series of gyroscopes, very similar to those contained in Dean Kamen's better-known Segway, to raise itself up on two wheels, allowing the person in the wheelchair to raise him or herself up to a standing person's eye level and make use of standard counters at places like the grocery store. The sturdy construction even allows users to navigate through rocky terrain or several inches of water. The iBot was discontinued for some years, but it's about to make a comeback. It's anticipated to be lighter, even more maneuverable, and one more step toward putting the world back within everyone's reach.

would usually need to be designed so that the pneumatic switch is activated by something external, perhaps by a mission model, releasing the stored air and powering the actuator. With VEX's ten-motor competition limit, running out of slots is not likely, but certainly not impossible. Additionally, pneumatics give VEX manipulators a stronger and faster response than traditional motors, making it easier to manipulate large game pieces, or several at once. However, there is often a wildcard factor in using so much energy to perform simple actions quickly.

Pulling and Pushing

Many pushing attachments are passive, requiring no additional power. They are attached to the robot and use the robot's movement to accomplish the objective.

A pushing or pulling action is horizontal or vertical, so linear actuators may be an obvious choice. One would have to add a pushing device, or possibly a hook, for pulling, but the operation would be simple: just activate the attached motor. However, linear actuators are not standard with most robot kits. If a rotational actuator is used, then a gear system must be used to convert the rotational actuator's motion to linear motion. One way could be to have a round gear attached to the motor, operating on a rack gear. As the round gear turns, the rack gear will be pulled in or pushed out. You could change the speed and torque of your attachment by changing the size of the gears being used. The larger the gear, the faster the attachment would move. If it moves very fast, you might need to add a stop device so that the attachment does not extend too far.

Though pushing or pulling devices like these are tempting, there are a couple of major drawbacks. Pulling hooks like these do not articulate in other directions and so make for very specific uses. More important, when you design an actuator-powered attachment for pushing or pulling, it is very likely that you're designing a redundant system. Any adequately powered and tractioned drivetrain will function perfectly well to push and pull objects around the game field. Overdesigning is a very real hazard that will add weight, bring you closer to your motor limit, and dramatically increase the time spent programming,

A LEGO robot using its frame instead of a manipulator to push some game pieces and score some points.

troubleshooting, and maintaining your robot. Anything that can be made to multitask should be able to multitask.

Grabbing

Often, a robot needs to grab things and return them to base or score with them in some way, so grabbing attachments are necessary. They can be compared to hands, but until quite recently, they were very inefficient hands because of the difficulty of understanding how much force to use to grab something. Commercially available and experimental robots could not really pick up delicate items, as they were liable to squash them. A lot of work has been done on this recently. A breakthrough was made some years ago with the DEKA "Luke" arm, commissioned by the military as a prosthesis for those who had lost an arm. There is an astonishing video on YouTube showing a man who had lost both his arms twenty years before using this prosthesis to pick up a grape, without squashing it, and eat it. Another fascinating field is "soft robotics," where the gripping attachment, or even the whole robot, is made of soft or inflatable materials that can grip and move very delicate things, like eggs.

Grabbing attachments can be used either on their own or at the end of an arm of some type. Some grabbing attachments can be passive, using hooks or one-way doors to catch the object and prevent it from slipping out when the robot moves in the opposite direction.

Other grabbing attachments use motors as actuators, quite often using more than one motor to allow the robot more

degrees of freedom. Each direction in which independent motion can occur is called a degree of freedom. If the grabber can rotate, like a wrist rotates at the end of your arm, that is one degree of freedom. If the arm can move in and out, using a rack and pinion, for example, that would be another degree of freedom. Movement up and down would be another, as would movement from left to right. Although a human arm, with all its muscles and joints, has far more degrees of freedom than most robot arms, the robot may have a greater reach, or be able

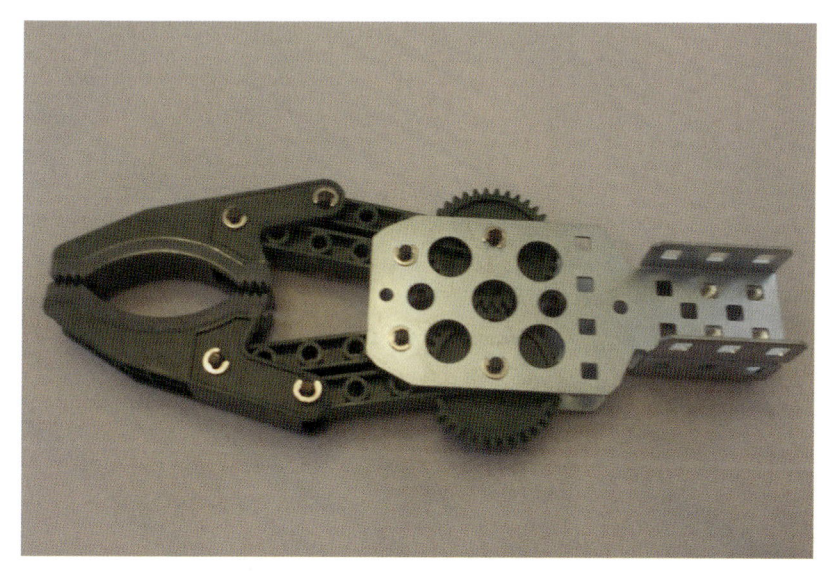

A standard VEX gripper using one actuator to move both pincers. The gears guiding the pincers spin in opposite directions.

to twist farther around than an arm can. When it comes to robot attachments, the simpler they are, the fewer degrees of freedom they have, the more robust they are likely to be.

The advantage of using a powered grabber over a passive attachment is that you can control the speed and power of your grabber, as well as the length of time it takes to close and to hold the cargo item.

A simple grabber is a trap, which can be designed in two ways. It could be a two-clawed grabber that snaps together around the object, rather than gripping it, moving the object home inside the cage thus formed. It could also be an actual cage, raised up using the rotary motor, then snapped down over the cargo item, which can be dragged home within the cage.

A claw is a very common grabber, using at least two pincers hinged at the same pivot point, one on either side of the object, coming together to grab hold of the object. A geared drive can be attached to the motor to make it possible for both pincers to be moved at once.

A similar gripper is a vise grip, which works in a similar way to an adjustable wrench. One arm of the grip cannot be moved, and the other arm moves left and right along a rack and pinion assembly using worm gears. This moves more evenly than the claw grip and can be controlled more precisely, so it

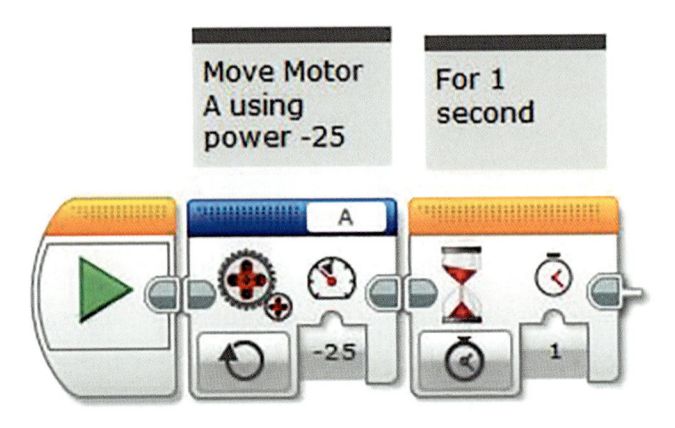

This program uses a Wait Time block to limit the movement of a motor.

is better for gripping more delicate items. With both a LEGO claw and a LEGO vise grip, or even an arm, you may find that your program requires the attachment to move farther than is physically possible. The program won't proceed to the next command until the movement command has been completed. You can either change the number of rotations or degrees in your program, or, more effectively, you can program the attachment to move for a specified time. Once the time is up, the program will continue no matter what the attachment is doing.

A very popular VEX attachment for grabbing items such as small balls is the roller claw. This uses pairs of rotating wheels or rollers, turning inward, to suck the balls from the game area into the robot's arms. These are very effective and don't require careful positioning of the robot, as once the rollers touch the ball or other item, friction will drag the ball toward the robot and between the rollers.

An engineer tests a robotic arm to determine how many degrees of freedom it possesses.

4 Help, My Robot Can't Get Up

What do you do when your robot has a problem? Do what engineers do. They use the **Engineering Design Process**.

If you look this up, you will find many different diagrams showing the Engineering Design Process, but they all describe a similar way of approaching a problem.

1. **Decide what the problem really is.** If your task is for the robot to throw a ball into a goal, is the problem that the robot is not getting the ball into the goal, or is it that it is not reaching the correct position in front of the goal?
2. **Brainstorm.** As a team, write down whatever ideas pop into your head about how to solve the problem. Don't judge ideas. If a teammate suggests using balls made of antigravity material, don't jeer—write it down. It might trigger an idea about changing the ball's trajectory that could solve the problem. If you

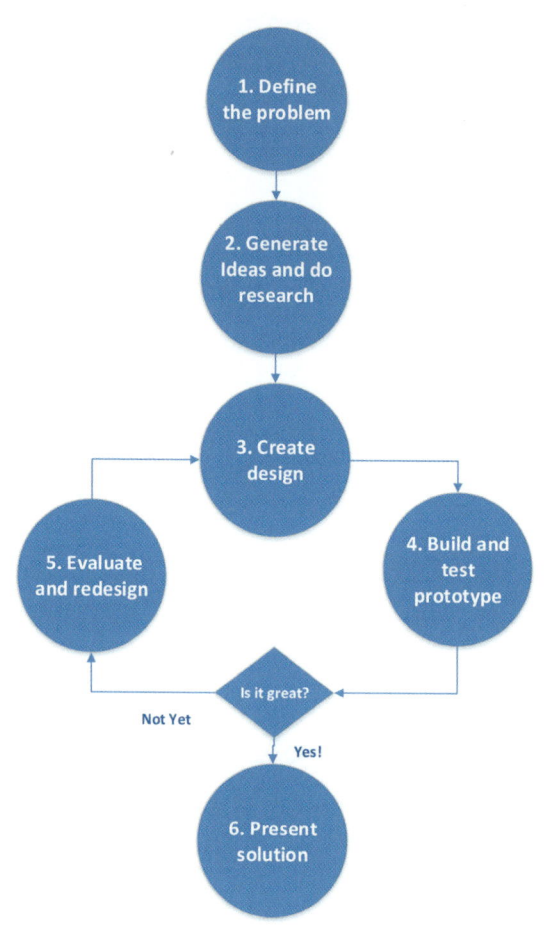

```
         ┌─────────────┐
         │ 1. Define   │
         │ the problem │
         └──────┬──────┘
                ↓
         ┌─────────────┐
         │ 2. Generate │
         │ Ideas and do│
         │  research   │
         └──────┬──────┘
                ↓
         ┌─────────────┐
    ┌───→│ 3. Create   │───┐
    │    │   design    │   │
    │    └─────────────┘   ↓
┌───┴────────┐       ┌─────────────┐
│5. Evaluate │       │4. Build and │
│and redesign│       │    test     │
└─────▲──────┘       │  prototype  │
      │              └──────┬──────┘
      │    ◇ Is it great? ◇ │
      └──── Not Yet         │
                   Yes! ↓
            ┌─────────────┐
            │ 6. Present  │
            │  solution   │
            └─────────────┘
```

The Engineering Design Process starts with an idea and ends with a working object.

find that your thoughts end up with your redefining the problem, loop back to step 1. If not, start judging the brainstorming results. If you have no good ideas, start over.

3. **Choose the best of these ideas**. As a team, decide on one idea to follow up. Decide in detail how you will make your idea work. If you need to build an attachment, draw a design. If you need to write a

program, make a **flowchart** or write an algorithm. If you can't come up with anything sensible, loop back to step 1 or 2.
4. **Construct and test your prototype**. Build your attachment, write your program, and use them.
5. **Evaluate and redesign**. Many diagrams of the Engineering Design Process either form a circle, or have a circle showing that steps 3, 4, and 5 can be repeated until you find a working solution.
6. **Present solution**. Goooooooooooaaaaaaaaaaaal!

This is a good way to go about designing and building your robot.

Understanding the Task

The first step in the Engineering Design Process is to understand your problem. When analyzing the use of actuators in robots, you need to be familiar with the following concepts:

Speed is the rate at which something moves or travels, or how far it will move in a unit of time—for example, centimeters per second. If you multiply the circumference of the robot wheel by the number of rotations per minute, you will get the distance your robot travels in a minute.

Acceleration is an increase in the speed of an object. The higher the acceleration, the faster the speed will increase. If the speed is constant, the robot is not accelerating.

Rotational speed is a measure of how fast something is circling around its axis. It is usually measured in revolutions per minute (rpm) or degrees per second. Different types of motors will have different rotational speeds.

Force is a push or pull on an object. Movement and acceleration are caused by forces. Forces can also cause a change of direction or shape. Robots accelerate because their wheels exert forces on the floor. Force is measured in newtons (N).

Work is done on an object when the force applied to it causes it to move. The formula for work is Work = Force × Distance, so you can see that if there is no movement, no work has been done.

Power is the rate that work is done upon an object. The term "rate" shows that power is a time-based quantity, so it means how fast the work is done.

Torque is a twisting or spinning force, a force directed in a circle. The word is also used as a measure of motor power.

How Do I Choose Which Motor to Use?

If you are using a LEGO Mindstorms EV3 robot, you have a choice between the large motor and the medium motor. Both motors have an internal gear train, gearing the motors down, but to a different extent.

The large motor runs at 160–170 rpm, and has a running torque of 20 newton centimeters (about 28 ounce-inches), so it is slower but more powerful. The rotation speed depends mainly on the power level being used, though the amount of load on the motor also has an effect. Pieces are attached via an axle inserted right through the motor, perpendicular to the length of the motor, resulting in a firm base. The design of the motor makes it easy to attach a gear train. The large motors are generally used to run the wheels or whatever is making the robot move. They should be used to power attachments.

The medium motor is smaller, and has a front axle hub. It rotates at 240–250 rpm, and has a running torque of 8 newton centimeters (about 11 ounce-inches), so it is faster but less powerful than the larger motor. The use of the front axle hub means that you have to build a support structure for attachments using the medium motor. This motor has a planetary gear system. A central "sun" gear and an outer ring gear rotate around the same center, with three small "planet" gears on a carrier. The three planet gears all engage with the sun gear, and they mesh with the inside of the outer ring gear, rather than the outside. This creates a very sturdy gear system and means that the medium motor has very little backlash, making this the right motor to use for low-torque, precise attachments.

The less bulky design of the medium motor may also simplify the design and building of the robot.

In summary, the EV3 large motor would generally be used to run the wheels, as well as powering attachments that need to lift heavier items. The EV3 medium motor can be used to drive the wheels, though this might require some creative programming and design (as you only have one medium motor in a kit), but would generally be used to power attachments that need greater speed and less torque.

With VEX, there are a few notable differences. The lack of planetary gears stands out because the VEX motor does not transform the direction of motion; the final output is parallel to the DC motor's output. As you saw when we looked inside before, the gears in the continuous VEX motor can be changed out to adjust the torque up or down. The two options are called the "high speed" and the "turbo gear" options. It is worth noting that these options are not able to be used for the VEX servo motor. However, you can approximate a servo by using a continuous motor, along with your own sensor or encoder. The last real limit on VEX only tangentially involves the actuators—how they communicate with the brain. The VEX Cortex, its microcontroller, has eight three-wire ports, but only two ports for two-wire motors. Therefore, if you want to use additional two-wire motors, you must use the VEX Motor Controller 29. The motor controller allows you to talk to a two-wire system as if it were a three-wire, and also to plug it into the three-wire ports.

There are companies that sell other motors that are compatible with VEX and EV3. Check the specifications of each if you need something the motors in the kit are not giving you.

Engineering the System

You need to be able to make the link between what a robot's actuators can do and the action a robot needs to take to accomplish its task. If you're having trouble figuring out the action a robot needs to take, there's a good chance that the task is not a simple one. By that, we mean there are multiple actions that need to happen in order to complete the task. With robots, you must go through tasks a simple, small step at a time. We tend to think of "taking a sip from a cup" as a simple action. However, it involves grasping the cup or handle, bringing it to your lips, pursing your lips, and rotating your wrist as you sip. Your wrist alone has three degrees of freedom—two more than our robot actuators. So for our purposes, it makes a lot of sense to break the task down as far as it can go. Divide each task—whether it is grasping, lifting, or pushing—into bite-sized instructions. For the task of grabbing, for instance, a robot needs to take more than a few actions. The actions would be positioning the wheels so the arm can reach, positioning the arm so the manipulator can reach, closing the manipulator to grasp the object, and sometimes even lifting the object clear of the surface it rested on. Ideally, at least for simplicity's sake, a single actuator can accomplish an action. In the previous example, the robot would be using at least four actuators—two for the drivetrain, one for the arm, and one for the manipulator. If having just the one is not feasible and it requires two actuators, then care must be taken to make sure they work in tandem and do not interfere with each other. The easiest and most common way that can be done is by reflecting oppositely placed motors, which you'll hear about momentarily.

Accuracy Versus Precision

When programming a robot to do a mission or complete a task, you need the robot to be both accurate and precise. Often, due to design, there is some give and take between which is prioritized, although having both is ideal. With many robots that can drive to a specific point and trigger a mechanism, accuracy is preferable. However, when trying to score goals or dealing with a remote-controlled robot, precision is more important. Knowing which matters more to you is a necessary step when choosing your manipulators and the actuators that power them. What is the difference between these words?

Accuracy means that the robot is on the exact spot that it should be, or its measurement is the same as or close to the real-life value. If your robot was supposed to stop in the red circle, and ended up on the edge of the circle on four separate runs, we could say that the robot was fairly accurate, but not very precise.

Precision means that the outcomes or measurements are very close to each other. So if the robot ended up in the same place, even if it missed the circle, on four straight runs, the results would be very precise but not accurate.

For example, when my class was doing robot "basketball," one team scored baskets with every throw, so they were both accurate and precise. Another team scored with about half their throws, so they were fairly accurate, but not precise.

A robot that needs to throw a ball and hit a target to score points requires precision more than accuracy.

Ouch! Common Issues

There are a number of issues that stand out as overwhelmingly common when it comes to building robots. Having a solid understanding of how to diagnose and repair issues is imperative, especially during competition seasons—or worse, a competition. Here are a few things to keep an eye out for, and some solid methods to troubleshoot pretty much anything.

1. My servo is wrong! My sensor is wrong! The robot is disobeying!
All of these, 99 percent of the time, have the same answer: no it isn't. You've probably been told this before, but I have some news to break to you. Computers are really very dopey. Despite the awesome graphics on your game console, you've probably seen anything from airplanes to people get stuck in the walls. Computers follow our instructions to a fault. If the robot is not

TROUBLESHOOTING TIP

If you're having multiple issues, break into teams and look at them separately. Is your group in disagreement over what's causing the issue? Always start with the least invasive possible solution! Quick motor swaps are sometimes easier than software changes, and software changes are often easier than tearing down a drivetrain. Choose what works for you, and go from there!

doing what you want it to, you probably didn't tell it the right thing. It's especially important to keep this in mind with your servos and encoders. The first thing to do is make sure that you're giving it the input it needs to do what you want. It is possible for those components to fail, but it is quite uncommon.

You can test servos by rotating them all the way in either direction—see if they have the range they should. Otherwise, the safest and quickest bet to check if it's your actuators or your code is to swap out one of the motors. If it continues to misbehave, you can change which port it is connected to. This will let you know if it's a problem with the port on the Cortex. Don't forget to swap out extension cables; they can have bad connections too! If it's none of those, most likely it's due to either your code or wonky design. Run your motor slowly and see if you can see any play from the axle, or movement in the body of the robot, that could be altering its output.

If the robot is not going the correct distance, you may find that you need to reset the rotation sensors to zero, as they may have started counting up from the previous setting.

Change one thing at a time to isolate the problem. If you know the problem, you can learn from your mistakes!

2. My robot is going backward!

You must remember that the robot does not have any specific orientation. Movement forward is programmed as positive power, and movement backward as negative power, but what this means is that the motor spins in one direction if the power is positive and in the other direction if the power is negative. You should be aware of the orientation of the motors in your build; if they are mounted one way, positive power will make the robot move in the direction that you have called forward because of your design, but if they are mounted upside down, your robot will move what you call backward.

If your robot is going backward, you do not need to rebuild. Just remember to use negative power values to move forward and positive power values for backward. If you are using VEX and ROBOTC, you can set a motor to be "reflected"—it will read the opposite of the value you put into the program, so that you can use intuitively obvious numbers without having to think about it.

3. The robot isn't doing what I told it to!

This is a very common complaint, to which my answer always is … Yes it is.

Robots like our school robots are incredibly stupid. Although most robots are considered to follow the expression, "Sense, Plan, Act," in fact, the planning is done by the programmer. All the robot can do is follow the program exactly. This means that the robot is doing exactly what you told it to do. The problem is that you have not given it the correct instructions.

How do we know what instructions to give? It's a good idea to plan them beforehand. Look carefully at the mission or objective, and break it down into small parts. You can start by breaking it down into obvious divisions, then try to break each of those divisions down further. What you end up with is an algorithm. The classic example of this is looking at how you make a peanut butter and jelly sandwich. One could say:

Every step must be carefully plotted in order to program a robot to spread peanut butter and jelly on separate pieces of bread.

1. Take two slices of bread
2. Get a jar of peanut butter
3. Put peanut butter on one slice of bread
4. Put jelly on the other slice of bread
5. Clap the two slices together

This would not be clear enough for your very stupid robot, so you would have to break each step down further. To break down "take two slices of bread":

1. Place the bread on the counter
2. Undo the tie holding the bag closed
3. Place the tie next to the bag

4. Insert hand into the bag
5. Grasp slice of bread
6. Remove slice of bread
7. Place bread on plate (making the assumption that a plate is nearby and the robot knows this)
8. Release slice of bread
9. Repeat steps 4 through 8

If necessary, each of these steps could be broken down even further. Here is how you break down "place the bread on the counter" (I keep my bread on a shelf):

COPY THAT

Once you have told the robot how to place the bread on the counter, you can create a function, subroutine, or MyBlock called *getBread*, which you can call whenever you want the robot to execute this action. It makes life easier when you don't have to type all those tedious instructions over and over again.

1. Open hand
2. Extend arm
3. Grasp bag of bread
4. Lift bread
5. Move bread back 15 cm
6. Lower bread 30 cm (*or* until it touches the counter)
7. Release bread

You later need to be clear, as a further example, that the peanut butter is on one side of the bread only, the same thickness over the whole surface of the bread, and that the peanut butter side of one slice should touch the jelly side of the other slice when you put them together. And you might tell it to use a knife to spread the peanut butter. It won't know that otherwise. That is the kind of detail that you need to use with a robot.

This breakdown can be tedious, but it is necessary. A good way to do this is to draw a flowchart. Some people consider flowcharts old-fashioned, but I teach them because it is a very clear way of analyzing a problem, and also because I realized that each block on the flowchart corresponded to an EV3 programming block. Once you have a good flowchart, your program is easy to write. Flowcharts are explained in more detail below.

Another suggestion if you are, for example, doing a table mission with your robot, is to move the robot over the table, doing a physical mock-up of the mission, and use sticky notes on the table to describe what the robot should be doing at each point.

1. Robot moves forward 30 centimeters (or number of rotations)
2. Robot turns left 90 degrees
3. Robot moves forward 10 centimeters
4. Robot lifts arm 45 degrees

You can then convert the sticky notes into a set of step-by-step instructions, or into a flow chart.

It is absolutely essential, when programming, to write just a little bit of the program at a time, then to test that it works the way you want it to work, before you write the next section. It is extremely difficult to find a mistake in a long, untested program, as it could be in an unexpected place. If you write the program bit by bit, testing it each time until it works as far as you have gone, you have a good chance of being successful. Document your work. If you are using EV3, write comments above each block or set of blocks, explaining what the robot is doing. This makes it so much easier to analyze or modify later. If you are writing in ROBOTC, add comments before each function or section explaining what the objective is. Your teammates will be happy. Your teacher will be happy.

If you have written a program, and it is not functioning as planned, another option is Rubber Ducky Debugging. Before you laugh, some professional programmers actually do this—they have a rubber ducky at their computer, and if they are having a problem, they explain their program in detail to the rubber duck, as though it were a colleague. Very often, as they are explaining their code, they understand what they have done wrong. It works for math problems as well: if you have trouble with a math problem, explain what you are doing to someone else (or the duck), and the chances are you will figure out your mistake.

Problem Solving

Breaking Down Tasks

Your first task when problem solving is to understand the problem. A good way to do this is to break the task down into smaller tasks. This brings us back to flowcharts. A flowchart is a diagram that shows the step-by-step progression through a procedure using connecting lines and a set of conventional symbols. The oval shape is used to indicate the start or end of a program. The rectangle represents a process or an action—for example, move forward. The parallelogram represents inputs to and outputs from the system. In a robot, these inputs might be from a sensor, such as the level of light observed. The rhombus represents a decision or a question. The answer to the question decides which path the program will take, following the answer on one of the arrows.

You can see that there is a loop in the flow chart on page 84, where the robot repeats the behaviors of moving forward and checking for red over and over.

Another way of planning programs is to use **pseudo-code**. This simply means to write down what the robot needs to do in English, giving all necessary details. To detail the program that appears on page 85, we could write:

1. Move forward
2. Check the floor color
3. If the floor color is red

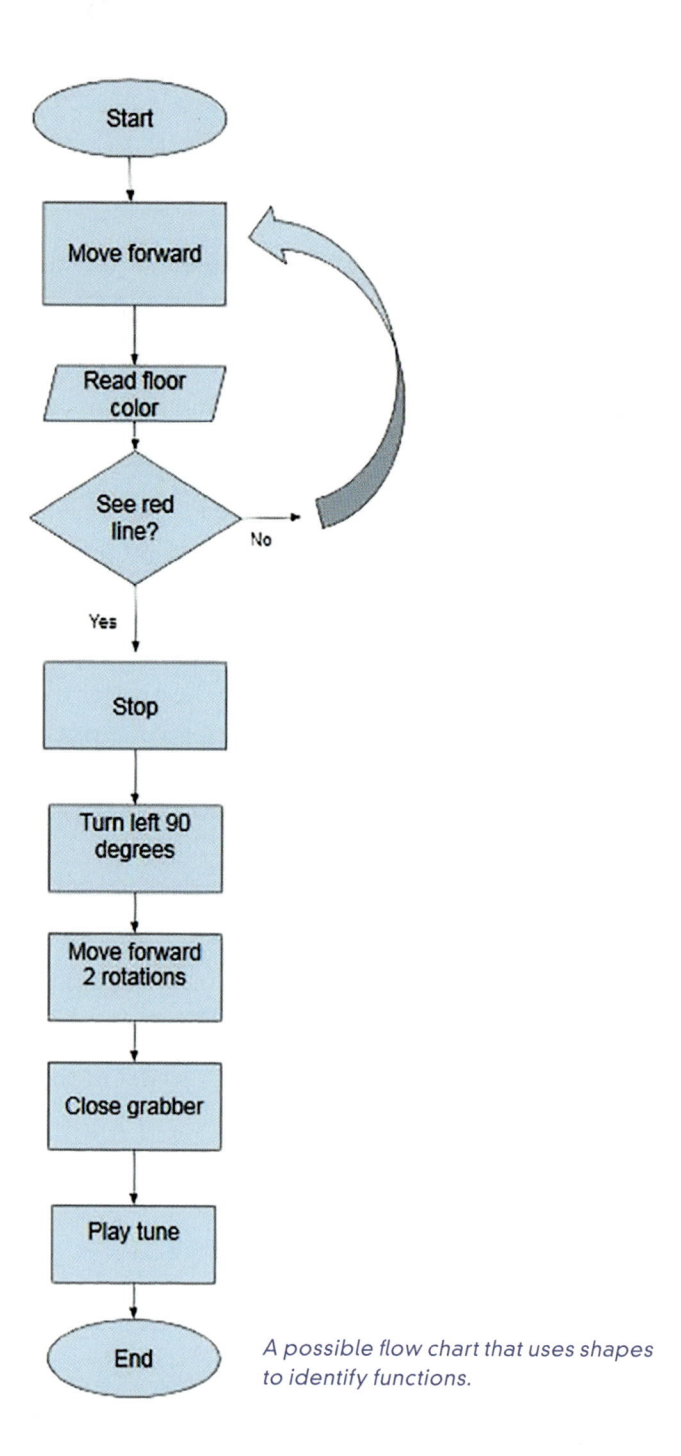

```
Start
  ↓
Move forward ←─┐
  ↓            │
Read floor     │
color          │
  ↓            │
See red line? ─┤ No
  │ Yes
  ↓
Stop
  ↓
Turn left 90
degrees
  ↓
Move forward
2 rotations
  ↓
Close grabber
  ↓
Play tune
  ↓
End
```

A possible flow chart that uses shapes to identify functions.

4. Stop

5. Turn 90 degrees to the left

6. Move forward two rotations

7. Close the grabber

8. Play a tune of triumph

If you were programming this in ROBOTC, you would need to use a loop so that the robot would keep looking for its objective. EV3 has a Wait Block that makes the loop unnecessary in this particular case. The corresponding program for the EV3 would be:

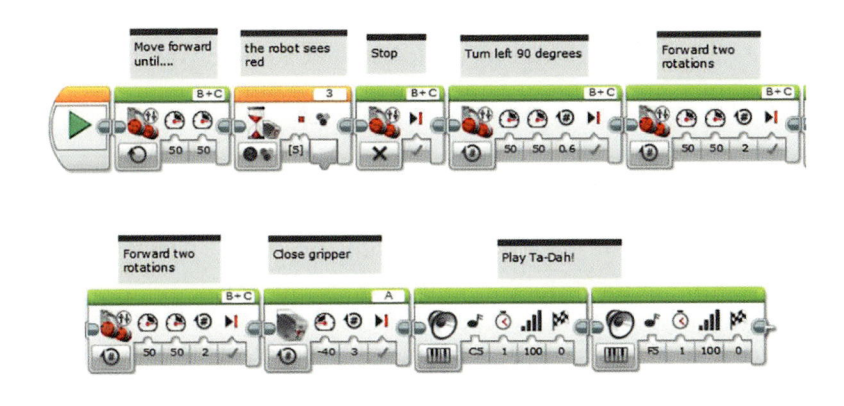

A sample EV3 program using a wait block instead of a loop.

It really doesn't matter how you write it, as long as you plan the program before you write it. The more professional way—as many developers use this structure to plan their code—is to write your pseudo-code like you were already talking to the robot. This allows you to speak the robot's language before you actually dive in and can make it easier to translate to actual code. For example, it would help you to remember to use if-then statements or brackets to outline loops. Some professionals use flowcharts, too. It's up to you which one you use. Ideally, it should be whichever one makes more sense to your own brain. The benefit to using pseudo-code is that it can be formatted like code, allowing you to see the connections you are making without struggling with syntax.

Troubleshooting Actuators

You're having problems with your actuators, and you understand flowcharts. Here are examples of a series of steps to place in a flowchart and to work through to help you identify your problem:

Do both motors turn?
    If not, are the cables plugged into the output ports?
        If not, plug them in.
    If they are, are the ports the same as in the program?
        If not, make sure the program ports match the cable ports.
    If they are, is the battery charged?
        If not, charge or change the battery.
    If it is, switch out motors and controllers one by one to isolate the problem.

If the motors turn, is the robot going in the right direction?
    If not, is the power value negative?

If it is, change the program power value to a positive number.

If not, have your motors been installed upside down?

If they have, change the program power value to a negative number.

Teamwork and Strategies

You will always be part of a team when working with a robot. Often there is friction between team members, as some people might feel that one team member is dominating and not allowing the others to do their share of the work. The team can set up structures to minimize problems.

The team needs to make sure that each team member has about the same amount of work to do. Hogs should not grab additional work because they feel no one can do it as well as they can, and Logs should not sit back and watch. Each team member's strengths and weaknesses should be considered. Every student should learn. If the best builder does all the building, and the best programmer does all the programming, it may look as though the team is pursuing a winning strategy, but in fact you are not working as a team. If some students have little experience, apprentice them to an experienced student so that they can develop expertise. Apprentices can switch between topics, at previously decided intervals, so that they get a balanced training.

Because no programming should be done before the robot has been built, the whole team can be involved in design and building. It's a good idea to use the Engineering Design

Process to develop a robot design—understand the problem, brainstorm, develop two or three design ideas, and choose one to build. Equally, the whole team can be involved in developing an algorithm (a set of step-by-step instructions) or flowchart for the programs. If you find that the team is splitting into definite groups with different responsibilities, it's a good idea to have a five-minute team meeting at the end of each session so that everyone knows what is going on. If you are a competition team, remember that the judges can pick on someone randomly to answer a question, and they don't appreciate your not being able to answer because it was someone else's job.

Before you can decide how to allocate the work, you need to know what work needs to be done.

You could use a Work Breakdown Structure for this, in which each task for each step in the Engineering Design Process is outlined. Once the team knows what needs to be done, the members can create a chart showing who is responsible for which tasks and when those tasks need to be done.

Out-of-This-World Solution

Have you ever had an itch on your back you just couldn't reach? Many of us use a pen or pencil to reach the itch—other people will scratch their back on a vertical surface like a bear. But what do you do when a space station's manipulator needs to reach somewhere it can't? Building manipulators—arms or extensions of your robot that move or grab game parts—is one of the great joys of robotics, as they are flashier and rather more fun than

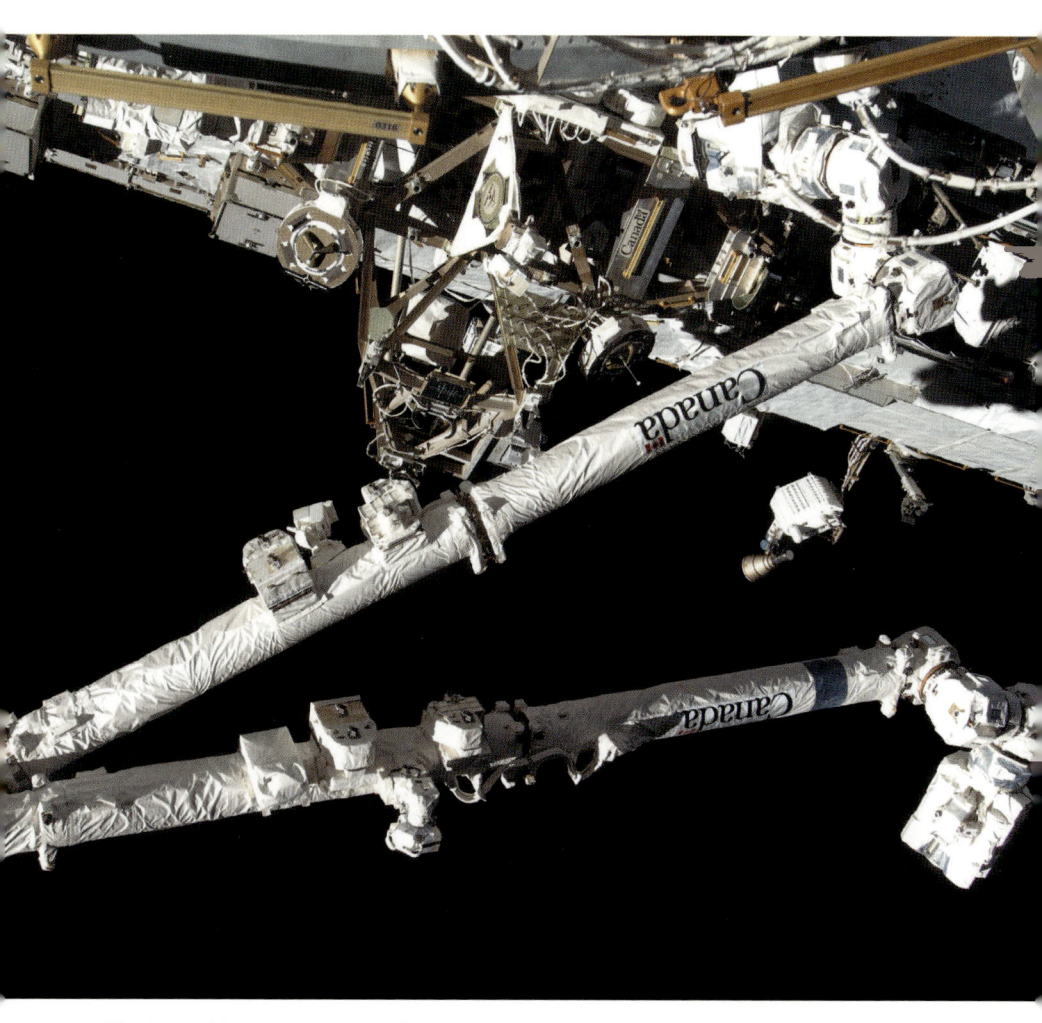

The incredible design of the Space Station Remote Manipulator System allows for access almost anywhere on the International Space Station.

your standard drivetrain. However, building manipulators can present great challenges. Nothing can illustrate that better than the history of the Canadarm2.

The original Canadarm was the primary manipulator used on the now-retired space shuttles. It was much like a human arm, with six degrees of freedom, with its reach additionally limited by its length. It could hold things out and bring them into the shuttle's cargo space, but its dexterity was about as limited as a human's. However, when the International Space Station came about, the Canadarm2 was commissioned to assist in building the space station, as well as to perform tasks after its completion. The second Canadarm is properly called the Space Station Remote Manipulator System. The name alone should give you an idea of how much it is expected to accomplish. A limited reach would have been unthinkable for this work, and deploying multiple manipulators would be just as impossible. Some serious out-of-the-box thinking was required, and sure enough, a unique solution was developed for the Canadarm2. The Canadarm2 has seven degrees of freedom and is symmetrical over its center point. Because of this, the manipulator can reach down and grab one of the many power nodes located on the ISS and connect itself to power and data through its end effector. An end effector is the part of your manipulator that interacts with the environment, like your hand does for you.

We don't have the benefit of this next part on our hands, though: the Canadarm2 can detach the "shoulder" it was using before, and use it as the new "hand." In doing so, it can swing itself along the ISS, much like a human using monkey bars. It has sensors that give these hands a sense of touch. With power nodes all over the space station, the arm has access to virtually every point on the exterior of the satellite. For repairs, docking assistance, and assisting with assembly, the versatility of the system is virtually indispensable. What's more, thanks in large part to the reduced forces in microgravity, the 1.8-ton (1.6-metric-ton) arm runs on about as much power as a hair drier. Working along with several other robotic manipulators, the Canadarm can intercept satellites, position parts for retrieval or placement when astronauts spacewalk, and even work with another robot named Dextre to perform certain repairs and external maintenance. Dextre has arms and tools to do some of the finer tasks, and it holds objects for Canadarm while it moves about the space station.

In a more down-to-earth twist, technologies pioneered for the Canadarm are now being used to remove otherwise inoperable tumors. While still remotely controlled by a surgeon, robotic arms can be much steadier and more precise than human surgeons, and they can work in more difficult spaces. One of the technologies being developed allows surgeons to operate on a patient in an MRI machine, which has never before been possible, a great reminder that solutions can come from anywhere, out of nowhere, or from outer space.

Different sized driver gears allow bicyclists to choose between power and speed depending on the terrain.

5 Moving Forward

You should now have a basic knowledge of how to use actuators in a robot. You still need to think about how you can improve the robot's performance. A robot's performance is decided by all aspects of the design, but the ability to move and act—to be athletic—is an uncompromising part of that. Hopefully, what you have learned in this book is enough to guide you through the difficult process of powering your robot. We're going to go over just a few more things, starting with how to move that power to the places you need.

We've talked about torque a lot throughout this book. It's important: this book is about power, and torque is very closely related to that. But you may have noticed that we've been generally referring to torque as a static property of rotational actuators. It goes up to a certain limit, and that's all there is. When you are choosing your actuators, that is generally adequate. However, torque can be manipulated through the use of gear trains. Gear trains are used for many reasons, but the biggest one is to transfer motion.

Two-Gear System

Driving Gear
36 Tooth

3 : 5 Gear Ratio

Driven Gear
60 Tooth

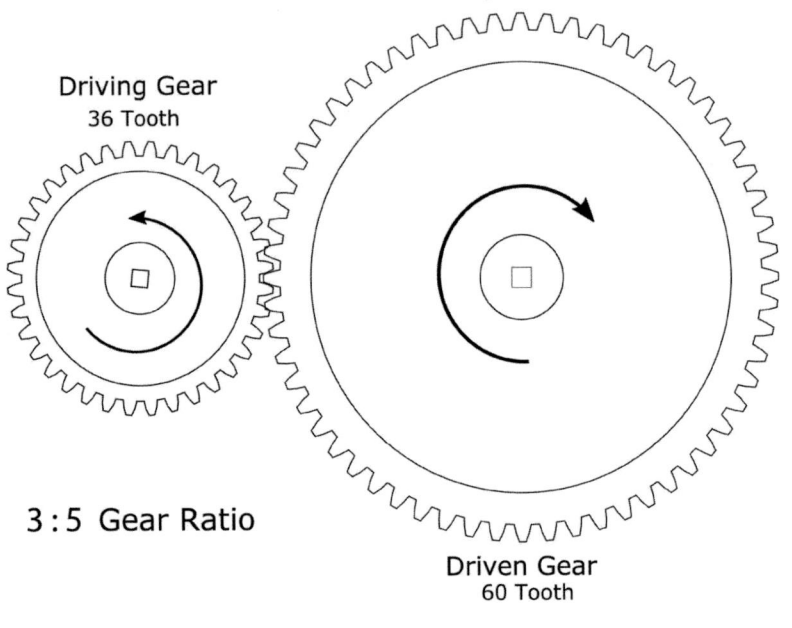

In a two-gear system, the gears turn in opposite diretions.

The above diagram shows a driving gear and a **driven gear**. The driving gear is always directly connected to the actuator, and it moves the rest of the gear train. The driven gear is located at the final output, in most cases, a wheel. Notice that the driven gear, or the gear that is being moved by another gear, is moving in the opposite direction of the driving gear, the gear that is providing the motion. There are two ways to offset this. One is to use a chain drive, or two gears (usually called sprockets when they interface with a chain) connected by a chain, just like your bicycle chain. This is a highly effective method because it can span relatively large distances, and more flexible distances, than a long set of gears. It also accomplishes this without introducing a lot of friction into the system. Don't forget, friction is not often our friend.

The other way this can be accomplished is by introducing an idler gear. Think back to the actuators we looked inside earlier in the book. The gears inside were shifting the rotary motion, rather than actually affecting the power. Gears used like this are called idler gears. They are very important when designing your drivetrain, not only because they are very useful, but because they are often the cause of breakdowns or malfunctions if the gear train is poorly designed. Idler gears are used to reverse motion. Since the idler gear sits between two working gears, motion is reversed twice. In the same way that two negatives make a positive, a three-gear system moves in the original direction. You can see both of these at work in the dual diagram on page 96.

Even better, the size of an idler gear can be determined by the amount of distance you have to span. Since we don't need to know the number of times it rotates to transfer motion, it does not affect our gear reduction. Gear reductions lower the speed of output, resulting in higher torque. It can be thought about very simply, especially since, with our standardized systems, it is very easy to know how many teeth a given gear has. Since all our gears that will mesh have the same size teeth, we can use the number of teeth as an approximation for the size of the gear. Fewer teeth means a smaller gear; more teeth mean a larger gear. Larger gears have more leverage; smaller gears complete a full rotation more quickly. This also means we can talk about rotation in terms of how many gears have moved, very much like a stepper motor.

Gear train with idler gear

Chain Drive

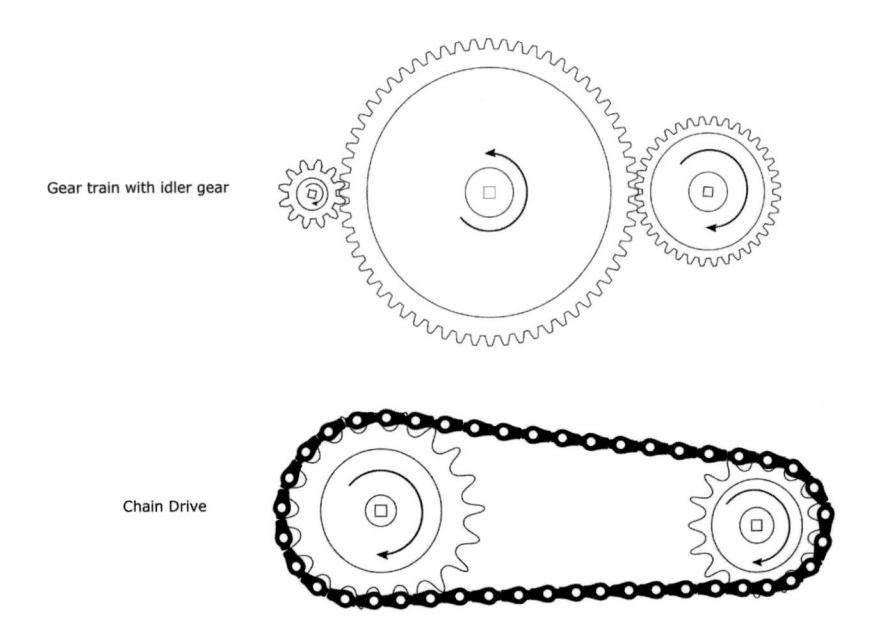

An idler gear (top) or a chain drive (bottom) allow the driving gear and the driven gear to turn in the same direction.

Let's imagine connecting a small, twelve-tooth gear to our motor, and a large, sixty-tooth gear to the wheel; a full rotation of our small driving gear only rotates the large gear twelve teeth, or one-fifth of a full rotation. Because it takes the larger gear longer to rotate once, it, and anything attached to it, moves at a slower speed. So if a wheel is attached to the larger driven gear, it will rotate more slowly than the motor, making the robot move at a slower speed. However, since the larger gear has more leverage, the slower rotation results in higher torque, and the robot will have more power, perhaps to pull a heavy load or to climb a steep hill. This relationship works both ways. If we reverse the gears, it only takes one-fifth of a large-gear rotation to make the twelve-tooth gear rotate 360 degrees. This translates to higher speed, but lower torque.

We express this relationship in terms of a ratio. The example on page 94 has a 36 to 60 **gear ratio**. It can be expressed as a fraction, 36/60, or in the more familiar form for ratios, 36:60. Either way, it can be reduced, just like you reduce a fraction. In this example, we end up with a 3:5 ratio. If the large gear comes first—that is, if it is the driving gear—the ratio would be 5:3. The smallest gears that are available with VEX are twelve-tooth gears, and the largest are eighty-four-tooth gears. That is a 1:7 ratio. What if you need a lower ratio? Delightfully, gear trains can be as complex as you like, and they have the wonderful benefit of being three dimensional. If you have two gears on the same axle, or fused together as we saw in the VEX motor's gearing sets, then you have a compound gear. Compound gears can be used to create an even greater gear reduction, resulting in more torque. We can stack a small gear on the same axle as our eighty-four-tooth gear. When we connect that small gear to an additional thirty-six-tooth gear, the small twelve-tooth gear and the thirty-six-tooth gear have a ratio of 1:3. To see the overall effect, you simply multiply the gear ratios. This is more easily done in the form of a fraction.

1/7 × 1/3 = 1/21.

So our final gear ratio is 1:21. Easy!

How does this really, truly change torque? The way these gears act together is, once again, by acting as a lever. Torque is effectively the force on or perpendicular to a lever, and speed is the distance the lever covers. If you look at the formula we provide in earlier chapters, you can see the relationship again. The longer the lever, the more torque is altered. The more teeth there are, the larger the gear, the wider the radius of the gear. Since the radius of the gear is effectively our lever, we can condense the whole thing to: the larger the gear, the more torque it needs to drive from the axle, and the less torque it takes to be driven from the teeth. We can reverse the process also, and think about increasing speed rather than increasing torque, simply by working in the opposite direction—a large driving gear on the motor to a small driven gear on the wheel.

When talking about speed, we can't forget angular velocity. You should already know that velocity is an object's speed in a given direction. Angular velocity is also known as rotational velocity, because angular velocity is the measure of speed over the distance of an angle. The formula is a little different than our others, because now we're talking about change over time.

Angular Velocity = Change in Position/Change in Time

Change in position is the distance an object moving rotationally covers, expressed as an angle. Change in time is simply time elapsed. Once again, the units follow the formula. Angular velocity would be expressed in degrees per second, or a similar unit. If you want to find how far your robot will travel in that time, you can use the circumference of your wheel, divided by 360 degrees, to approximate a distance traveled per degree. This is definitely only approximate: it does not take into account traction, friction, starting and stopping forces, or anything else that could lead it astray. But it will give you a good approximation, meaning you spend less time on trial and error and more time building something that works. You can even roughly calculate velocity with your encoders. If an encoder has two hundred increments, and it takes ten seconds to complete a rotation, the angular velocity is twenty increments per second.

All this theory makes it easy to understand what your robot will do once you have designed it, but designing it is often the hardest part. In this way, we are lucky to be using standardized systems. Using EV3 or VEX makes rapid prototyping easy, since not only do all pieces fit together easily, they work with our support beams to make spacing issues virtually disappear. This makes rapid prototyping very simple. In industry, supports for proposed gear trains usually have to be machined, often by hand, for testing.

For VEX and EV3, life is easy. Any gear train can be as long or as short as you need it to be. Whether your drivetrain has a nine-gear train (per side, of course, as you are most likely designing a symmetrical drivetrain) or a set of four gears with a compound element, they all function pretty much the same way.

There are many ways to experiment with the structure of your robot as well. Chain drives are ideal not only for your drivetrain, but for lifting mechanisms like forklifts. Chain drives also have the great advantage of using gear ratios in the same way as a standard gear train. A twelve-tooth sprocket driving a thirty-six-tooth sprocket still has a 1:3 gear ratio.

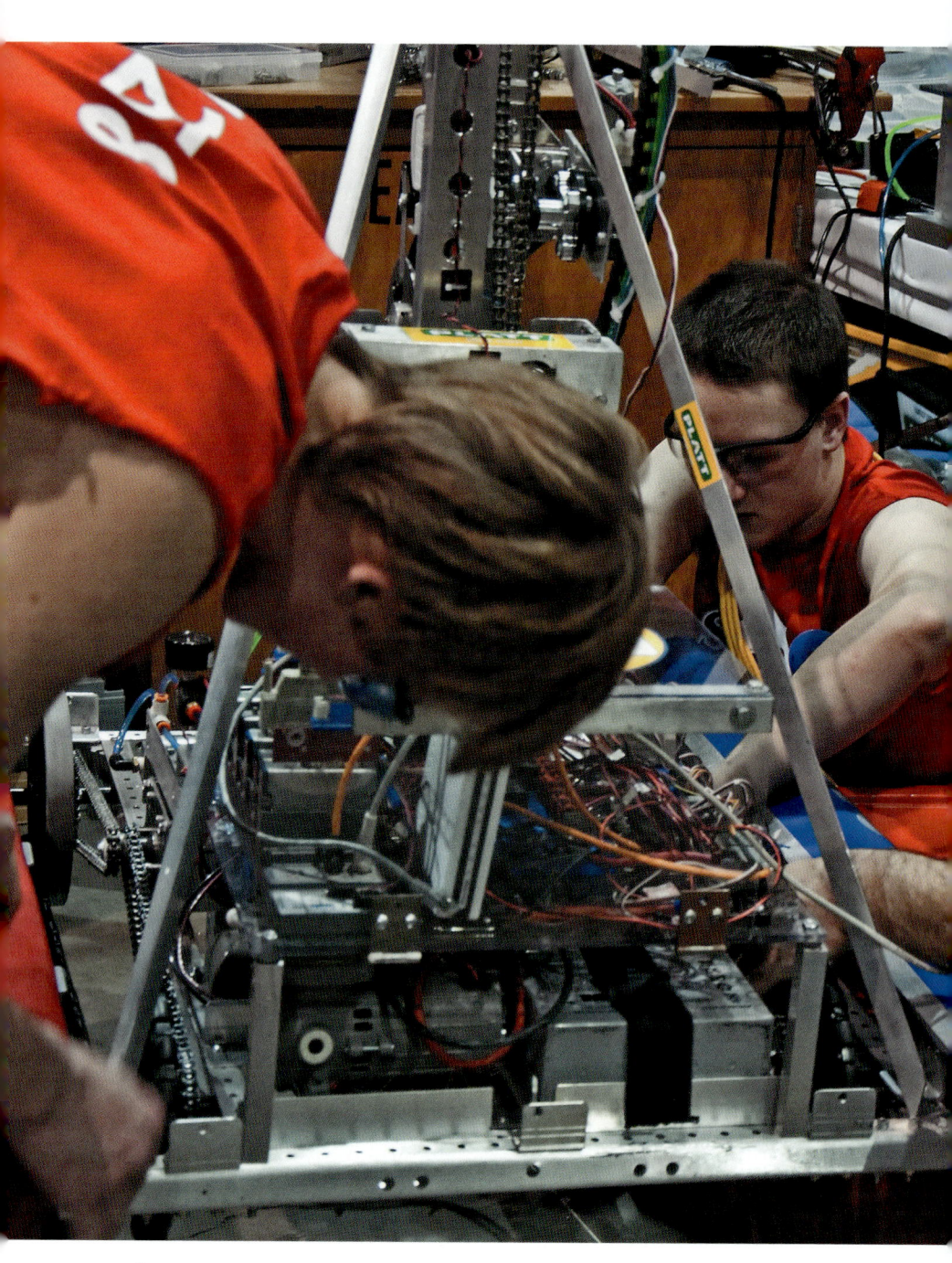

Competitors make adjustments to their robot, which uses separate chain drives to move and make turns.

Another neat trick to use for VEX is implementing miter gears, which allow you to transform motion at a 90-degree angle. Cam and followers can be used as anything from a release mechanism for game object storage to the "draw" on a slingshot mechanism. There are many options for you to experiment with.

Given the huge number of options, it can be daunting to know where to begin. The best way to get started is to look for mechanisms that spark your interest, whether in the real world or from past competitions. The sky's the limit switch, as long as you know how to read it, which leads us to our next topic.

Actuator Programming in ROBOTC

ROBOTC may look much more complicated than a drag-and-drop interface, but it is just as logical, which makes it quite easy to understand. Although it is quite fussy about indentations, the program will **indent** for you. Therefore, don't alter what it does for you.

It is important to spell correctly, and to match cases. If you use a capital letter instead of a lowercase letter, the program will not run, as the robot will not recognize the command. ROBOTC also color-codes words that it recognizes, which will help you notice typos and other errors.

The instructions that you give the robot are called **statements**. An example of a statement would be any command that you give the robot. The robot reads these commands in the order that you give them, generally ignoring spaces, tabs, and line breaks, except where they separate words in the statements.

Punctuation is key to programming. Each opening curly bracket must be matched with a closing curly bracket, each opening square bracket must be matched with a closing square bracket, and each statement in the program should end with a semicolon. Although statements do not need to be on separate lines as far as the robot is concerned, we usually type them on separate lines because it is easier for humans to read them that way. Different commands require different types of brackets or parentheses. This sounds complicated, but you will quickly get used to it.

Although we said that the robot reads statements in order, you can change the order using control structures such as loops and conditional statements.

Moving forward, let's consider a VEX robot using two motors, one controlling each wheel. Remember, earlier we showed that VEX motor power ranges between 127 and –127, so the power level of 63 shown below is about half the maximum power. As we discussed before, to move forward, we need to make both motors rotate at the same speed, in the same direction. We might guess that the following program would be appropriate:

```
taskmain ()
{
        motor[port3] = 63;
        motor[port2] = 63;
        wait1Msec(3000);
}
```

Sadly, if you run this program with a standard robot constructed like the one in the picture below, you will find that the robot turns in circles. Why?

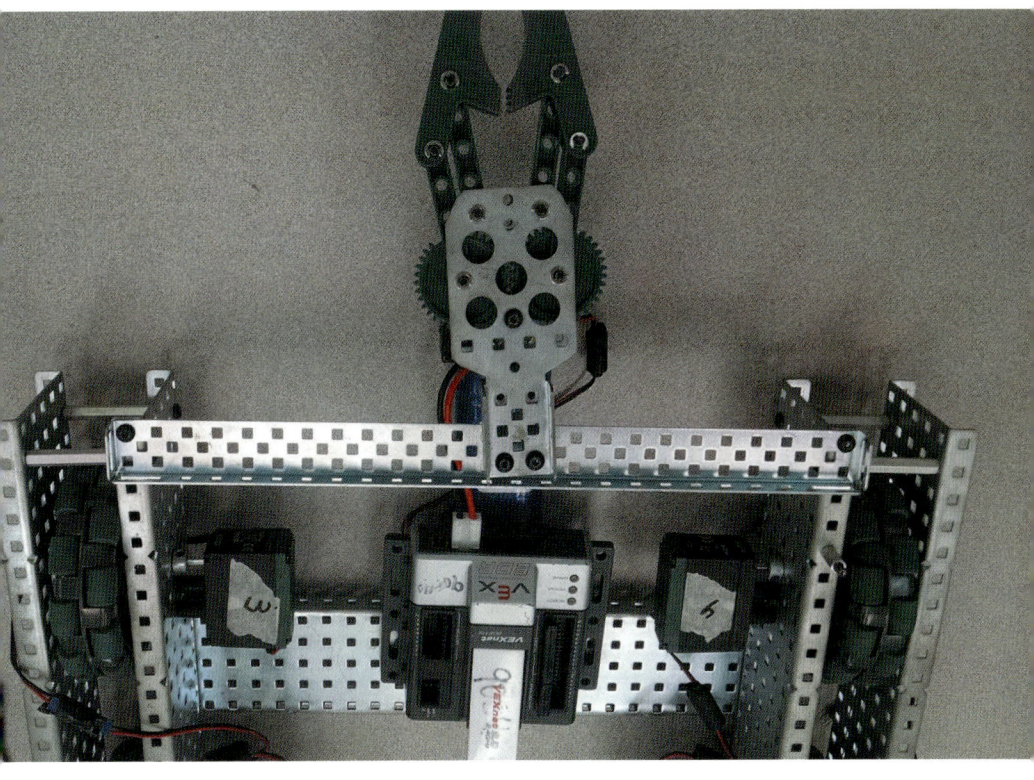

A mirror image installation will cause your robot to spin if both motors are programmed to move forward.

You will see, in the picture, that the two identical motors are installed in mirror image configurations. This means that the rotation direction that will move the robot forward in one motor will move it in the opposite direction for the other motor. We could, of course change the power value for motor two to −63, but it would be confusing, both when writing the program and when reading the program, to see that the motor powers had opposite signs with the robot moving forward. A solution is to use the *bMotorReflected* ROBOTC command, which will "reflect" the motor it refers to by 180 degrees, essentially changing the sign of the power value. This command causes the motor to hold that value for the rest of the program, unless the setting is changed. *bMotorReflected* is Boolean, meaning it can hold only two values: 0 or 1, False or True.

The following program will cause the robot to move forward for 3 seconds, then stop:

```
taskmain ()
{
    bMotorReflected[port2] = 1;
    motor[port3] = 63;
    motor[port2] = 63;
    wait1Msec(3000); // 3000 milliseconds = 3 seconds
}
```

Turns can be programmed the same way. To make the robot do a spin or point turn, turning on the spot, one wheel should move forward and one backward at the same speed, so the program would be:

```
taskmain ()
{
    bMotorReflected[port2] = 1;
    motor[port3] = 63;
    motor[port2] = -63;
    wait1Msec(3000);
}
```

The program for a pivot or swing turn, where one wheel is stopped and the other moves, could be:

```
taskmain ()
{
      bMotorReflected[port2] = 1;
      motor[port3] = 0;
      motor[port2] = 63;
      wait1Msec(3000);
}
```

These programs all use dead reckoning. This means that the robot's position is estimated using the relationship between distance, speed, and time. This type of calculation is very sensitive to a number of factors, so the robot's final position may vary depending on, for example, battery charge and floor composition. In order to move with greater precision, one needs to use encoders or servo motors.

Moving Using Motors with Encoders

Encoders, or rotation sensors, can be attached to the motors and are digital counting sensors, counting the number of times the axle, shaft, or wheel rotates. Most VEX encoders count 360 counts per revolution, counting up for forward movements, and down in the opposite direction. You can use the encoders to program the distance your robot moves in terms of rotations, as we discussed in chapter 3.

You do need to tell ROBOTC that you are using these rotation sensors and what ports they are plugged into before you can use them. ROBOTC has a Motors and Sensors Setup menu, which configures them automatically and will print a set of statements with their names at the beginning of the program.

It is essential to set the encoders to zero each time you use them to start measuring distance. If you omit this step, the encoders will start counting from whatever number they were set at when you started the movement, which will not give the results that you want. You will need to use a **loop statement** (*while*) so that the robot keeps checking the value of the encoder as it moves:

```
task main ()
/*
This program initializes the encoders to zero, then moves
forward at half speed, constantly reading the left encoder
value until left encoder reads 720, or two rotations. The robot
will stop once the reading is greater than or equal to 720.
*/
{
        SensorValue[rightEncoder] = 0;  //Set right encoder to 0
        SensorValue[leftEncoder] = 0;   //Set left encoder to 0
        bMotorReflected[port2] = 1;
        while{SensorValue[leftEncoder] < 720}; // Loop
        statement
                motor[rightMotor] = 63;
                motor[leftMotor] = 63;


}
```

Moving Using Servo Motors

When you set a regular motor to a value in ROBOTC, that value is used as a power setting, and the motor rotates continuously at that power. The servo behaves differently—it cannot rotate continuously and has a smaller range of motion. When you set a servo motor to a value in ROBOTC, the motor moves to that position and holds that position. For this reason, it is often used to power grippers and arms, or pan-and-tilt devices. The range of values for a servo motor is also from –127 to 127, where –127 is the farthest point it can reach in one direction and 127 is the farthest point it can reach in the other direction. Usually the VEX servo motors have a range of motion of 100 or 120 degrees.

To rotate a servo motor from the most backward point it can reach, to its central point, then to the other limit, you could write this program:

```
task main()
{
    motor[servoMotor] = -127;    // set servo fully backward
    wait1Msec(2000);             // wait for 2 seconds
    motor[servoMotor] = 0;       // set servo to middle of range
    wait1Msec(2000);             // wait for 2 seconds
    motor[servoMotor] = 127;     // set servo fully forward
    wait1Msec(2000);             // wait for 2 seconds
}
```

If you program a servo to move to a certain position, and your robot design has an obstacle that will prevent it from reaching this position, the servo motor will go on trying to reach that position. It does not know that it is physically impossible to get there. This is not only bad for the servo motor, but your program will not be able to proceed beyond that point, so you would have to change your program, substituting an accessible position for the servo motor. You can do this using guess and check, or you can use the debugging menu.

To finish, let's work on a program similar to the one in chapter 4. In pseudo-code:

1. Move forward two rotations
2. Turn 90 degrees to the left
3. Move forward two rotations
4. Close the grabber

```
#pragma config(I2C_Usage, I2C1, i2cSensors)
#pragma config(Sensor, I2C_1, rightIEM,
        sensorQuadEncoderOnI2CPort,  , AutoAssign)
#pragma config(Sensor, I2C_2, rightIEM,
        sensorQuadEncoderOnI2CPort,  , AutoAssign)
#pragma config(Motor, port1, rightMotor, tmotorVex269,
        openLoop, reversed, encoder, encoderPort,
        I2C_1, 1000)
#pragma config(Motor, port10, leftMotor, tmotorVex269,
        openLoop, encoder, encoder, encoderPort,
        I2C_2, 1000)
#pragma config(Motor, port9, servoMotor, tmotorNormal,
        openLoop)
```

```
//*!!Code automatically generated by 'ROBOTC'configuration
wizard  !!*//

/*++++++++++++++++++++| Notes |++++++++++++++++++++
Move Forward Turn Left and Grab
This program instructs your robot to move forward at half
power, turn left 90 degrees, move forward at half power,
then grab. There is a 2 second pause at the beginning of
the program.

Robot Model: Modified Clawbot
```

[I/O Port]	[Name]	[Type]	[Description]
Motor Port 1	rightMotor	269 Motor	Right side motor, reversed
Motor Port 10	leftMotor	269 Motor	Left side motor
I2C_1	rightIEM	Integrated Encoder	Encoder mounted on rightMotor
I2C_2	leftIEM	Integrated Encoder	Encoder mounted on leftMotor
Motor Port 9	servoMotor	Servo Motor	Motor on Grabber

```
------------------------------------------------------------------------*/
```

```
task main()
{
    wait1Msec(2000);

    //Clear the encoders on the left and right motors
    nMotorEncoder[rightMotor] = 0;
    nMotorEncoder[leftMotor] = 0;

    //Repeat the commands following until the right motor
    encoder reads > 720
    // This will move the robot forward 2 rotations
    while(nMotorEncoder[rightMotor] < 720;
    {
        //Move forward at half power
        motor[rightMotor] = 63;
        motor[leftMotor] = 63;
    }

    //Reset the encoders on the left and right motors
    nMotorEncoder[rightMotor] = 0;
    nMotorEncoder[leftMotor] = 0;

    //Repeat the commands following until the right motor
            encoder reads > 314
    //With the left motor off, and the right motor moving,
            the robot will turn left
    while(nMotorEncoder[rightMotor] <314);
    {
        //Move forward at half power to turn
            90 degrees
        motor[rightMotor] = 63;
        motor[leftMotor] = 0;
```

```
        }

        //Reset the encoders on the left and right motors
        nMotorEncoder[rightMotor] = 0;
        nMotorEncoder[leftMotor] = 0;

        //Repeat the commands following until the right motor
                encoder reads > 720
        //This will move the robot forward 2 rotations
        while(nMotorEncoder[rightMotor] < 720;
        {
                //Move forward at half power
                motor[rightMotor] = 63;
                motor[leftMotor] = 63;
        }

        //Set 'servoMotor' to position -127 (negative end of
                range)
        //Wait for 0.6 seconds (gives time for the servo to
                move)
        motor[servoMotor] = -127;
        wait1Msec(600);

}
```

    This has been a very basic summary of the commands used to control VEX actuators, but they should get you started and well on your way with programming your robot. Even school robots like LEGO Mindstorms need not be restricted to lessons and play. Shubham Banerjee, a seventh grader who

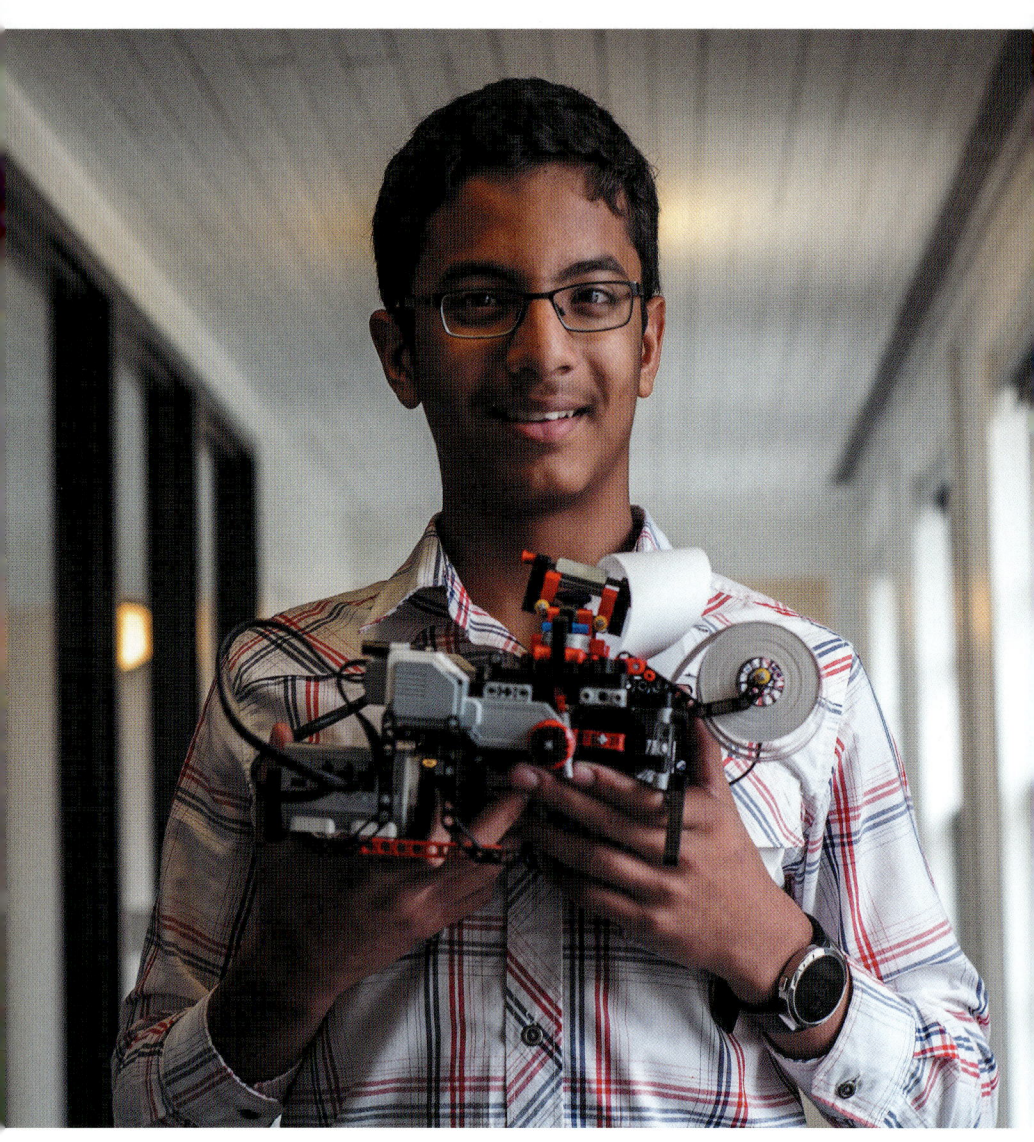

Shubham Banerjee with his braille printer built using a LEGO robot.

developed a low cost braille printer from a LEGO robot, inspired the LEGO Mindstorms RoboChallenge in 2014. LEGO asked six companies from the Seattle area to design and build LEGO robots that would help people. The winning design made it easier for parents of sick children to give the children their medication. The other designs included a robot to help take care of the house, a robot to pick up LEGO pieces, and a robot farmer. Even before that, a scientist in a lab in England felt that he and his peers were wasting their time meticulously dipping things in various solutions, slowly creating artificial bone for experiments, so he designed and built two LEGO robots to do the boring dipping for him. By learning about robots, you too can develop useful robots.

The power to do tasks is now firmly in your hands! Use it wisely.

Glossary

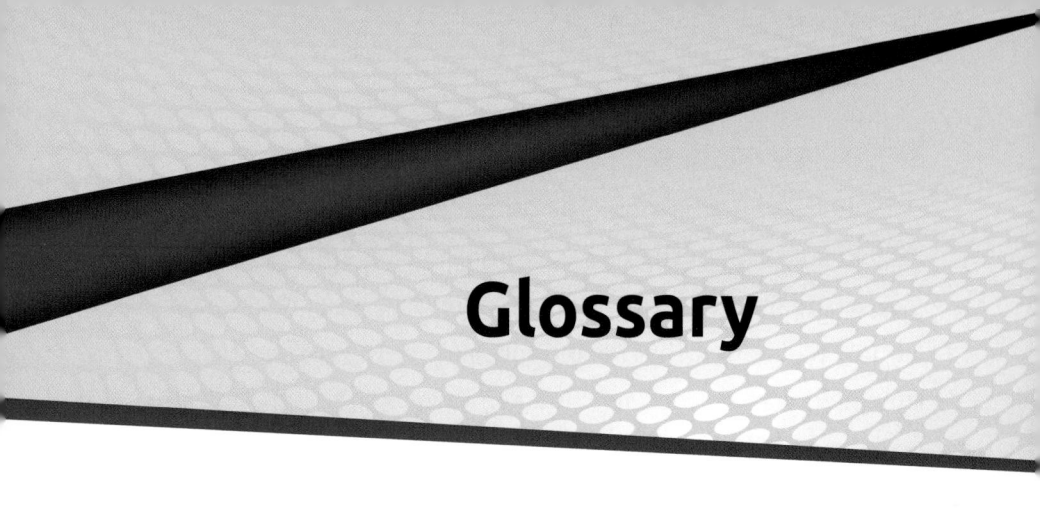

accuracy Close to the true or correct value.

AC motor An electric motor driven by alternating current.

actuator A mechanism that converts energy into movement or force.

cam A rotating or sliding piece in a mechanical linkage, often oval or pear-shaped.

circumference The distance around a circle.

curve turn A turn in which one wheel moves faster than the other, with both wheels moving in the same direction.

DC motor An electric motor driven by direct current.

degrees of freedom In a robot, the number of different directions in which movement can occur.

diameter A straight line through the center of a circle, connecting two points on the circumference.

differential gear A set of gears that allows two wheels that use the same power source to turn at different rates.

differential steering Steering in which movement is based on two separately driven wheels placed on either side of the body.

driven gear The driven gear is located at the gear train output, usually a wheel. It is moved by the driving gear.

drivetrain The components or gears between the motor and the wheels.

driving gear The driving gear is connected to the actuator and powers the gear train.

electromagnet A type of magnet in which the magnetic field is produce by an electric current.

encoder A device that monitors the position of a rotating shaft or axle.

Engineering Design Process A methodical approach to problem-solving in engineering.

flowchart A diagram showing the sequence of steps in a program.

forklift A device for lifting a heavy load vertically. It usually has two prongs, like a fork.

friction The force that makes it difficult for one object to slide over another.

gear A toothed wheel that meshes with another toothed wheel, changing speed, force, or direction of force.

gear ratio The ratio of the number of teeth on the driven gear to the number of teeth on the driving gear.

hydraulic Processes using water or another fluid under pressure.

indent A space left by moving a line of text, such as the first line in a paragraph or a line of coding, away from the left margin by a specified amount.

input Power, energy, or information provided to a machine or system.

lever A simple machine consisting of a rigid bar resting on a fulcrum.

linear actuator An actuator that creates motion in a straight line.

loop statement A flow control structure that causes a sequence of instructions to repeat.

magnetic field The area around a magnet in which the magnetic force acts.

mechanical advantage A measure of the drop in force required when you use a simple machine.

omniwheel A wheel on which the tread is composed of little rollers that allow the wheel to roll in many directions.

output The power or information leaving a system.

piezoelectric The ability to generate electricity when stressed or squeezed.

pinion A round gear that engages with a linear gear bar.

pivot turn A turn in which one wheel moves and the other wheel does not.

pneumatic Processes operated by air or gas under pressure.

point turn A turn taking up the least possible space, in which one wheel moves forward and the other backward at the same speed, so the vehicle spins around the center point of its axis.

power The rate at which work is performed or energy is converted.

power band The range of operating speeds at which a motor can work efficiently.

precision Outcomes or measurements that are accurate and exact; in machining it can mean getting the same result each time a tool is used.

pseudo-code A simple description of how a program works step by step, in English.

rack and pinion A mechanism in which a round gear engages with a bar or rack gear.

radius A straight line from the center of a circle to the circumference.

rotational motor A motor producing a rotary motion.

rotational speed The speed of an object rotating about an axis, often given in revolutions per minute (rpm).

scissor lift A surface that is raised or lowered by the closing or opening of crossed supports so it is pivoted like the two halves of a pair of scissors. This is driven by a linear actuator that has one end achored.

servo motor A motor which has a sensor for position feedback.

simple machines A mechanical device that makes work easier to do.

solenoid A cylindrical coil of wire acting as a magnet when carrying electric current.

solenoid valve An electromechanically operated valve, controlled by an electric current through a solenoid. It is used in linear actuators.

speed The rate at which something moves.

stall A condition when a motor stops rotating.

statement A command in a ROBOTC program.

stepper motor A motor that converts digital pulses into mechanical shaft rotations.

switch A device for making and breaking a connection in a circuit.

synchro drive A system where one motor moves all the wheels to affect direction, and another motor rotates all the wheels to produce motion.

system A set of connected parts forming a complex whole.

torque A force trying to rotate an object around an axis. The higher the torque, the higher the power exerted.

traction The grip of something on a surface.

turning scrub The frictional force between the wheels and the ground in a skid steer drive train when a wheel or wheels slide over the ground so the robot can turn.

wheelbase The distance between the two wheels of a robot or a vehicle.

worm gear A short revolving cylinder with a screw thread.

Further Information

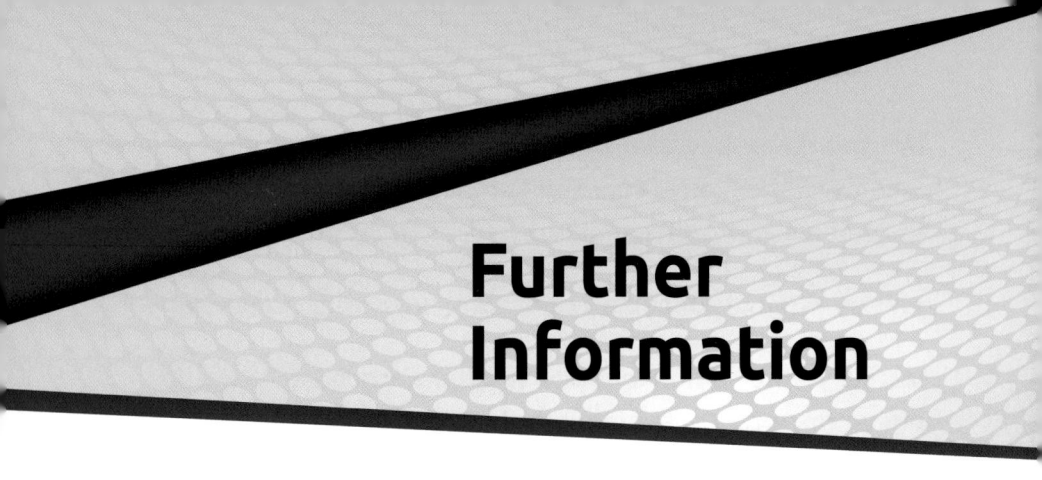

Competition Links

FIRST Robotics
www.firstinspires.org
Information about all the FIRST competitions can be found here, from elementary level through high school.

RoboCup Junior
www.robocupjunior.org.au/home
This Australian organization has interesting challenges that can be easily adapted for the classroom.

Robofest
www.robofest.net
A competition run by Lawrence Technological University. Even if you don't compete there, you can get good challenge ideas here.

Robotics Education and Competition Foundation
www.roboticseducation.org
This foundation runs a series of competitions for VEX IQ, and different tournaments for VEX EDR. Both middle schools and high schools can participate.

Technology Student Association
www.tsaweb.org
This is not strictly a robotics competition website but a general engineering site with robotics included.

Instructional Websites

Actuator Zone
www.actuatorzone.com/blog/technology/linear-actuators-robotics
A commercial site belonging to Progressive Automations, this site describes industrial applications of actuators in many different industries.

Building with LEGOs
spirit.mcs.uvawise.edu/LEGODocs/BuildingWithLEGOs.pdf
Good information on LEGO gears and driving mechanisms.

Carnegie Mellon Robotics Academy
education.rec.ri.cmu.edu
This site has very clear lessons and information covering both VEX and LEGO robots. There are also links to Virtual Worlds where you can program virtual robots, though these are not free.

Damien Kee's Technology in Education Page
www.damienkee.com
Dr. Damien Kee is an independent technology education expert, with several books to his credit and a website with a world of resources for NXT, EV3 and VEX IQ.

EV3 Lessons
ev3lessons.com
Very useful lesson videos and links.

Explain That Stuff!
www.explainthatstuff.com/hydraulics.html
More about hydraulics.

IKA Logic
www.ikalogic.com/small-robot-drive-trains/
Another commercial site, with a great page on small robot drivetrains.

LEGO Engineering
www.legoengineering.com
An essential website, full of inspiration.

LEGO Mindstorms Blog
www.thenxtstep.com
This keeps you up to date with LEGO Mindstorms developments and gives ideas and designs for amazing builds.

Milk and Cookies
milkandcookiesblog.com/teaching-with-lego-4
A blogger has several Teaching with LEGO posts, but this one about using LEGO pneumatics is great.

Nucleus Learning
www.nucleuslearning.com/lessonplan/teaching-hydraulics-and-pneumatics-unit-children
A great unit on hydraulics and pneumatics.

Philo's Home Page
www.philohome.com
A gold mine of interesting, eclectic information on both LEGO and VEX (though it's mostly VEX IQ, not the metal VEX EDR).

Physics for Kids—Electromagnetism and Electric Motors
www.ducksters.com/science/physics/electromagnetism_and_electric_motors.php
This gives easy-to-follow background on the physics behind motors.

Playing with Gears
scitechconnect.elsevier.com/wp-content/uploads/2013/09/Playing-with-Gears.pdf
A great description of how to use LEGO gears, from the book "Building Robots with LEGO Mindstorms NXT" by Mario Ferrari.

RobotShop—Making Sense of Actuators
www.robotshop.com/blog/en/how-to-make-a-robot-lesson-3-actuators-2-3703
A great summary of different actuators and what they do.

Teachengineering
www.teachengineering.org
This great site is full of fascinating ideas and lesson plans for kindergarten through high school teachers. Find in particular the entry on fluid power, and another about motors and rotation sensors.

VEX Machinations

www.vexrobotics.com/wiki/images/c/cb/VEX_Machinations-071108.pdf

Building instructions and pictures of many different robots. These are great for building confidence in your building techniques.

VEX Robotics Curriculum

curriculum.vexrobotics.com/curriculum

An entire set of lessons, pictures, and explanations.

Index

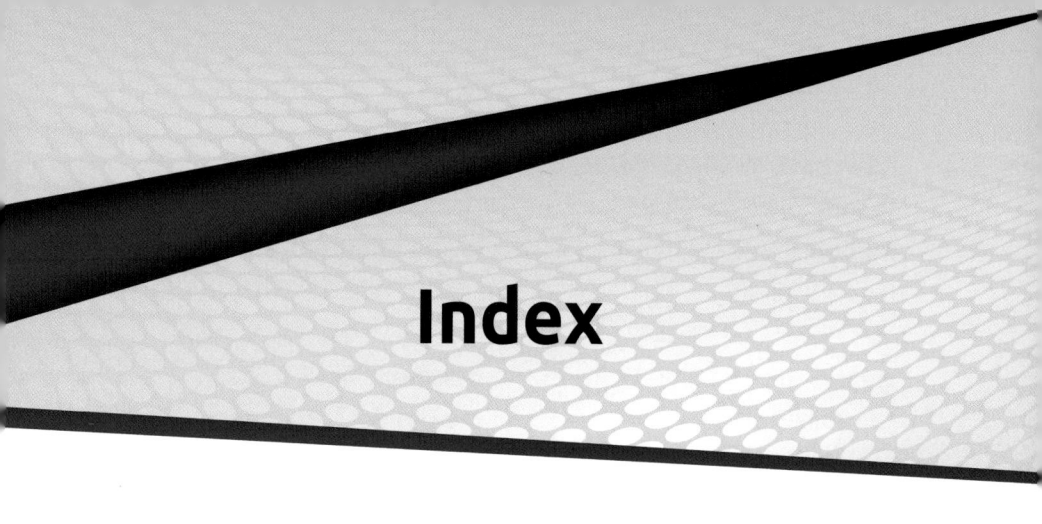

Page numbers in **boldface** are illustrations. Entries in **boldface** are glossary terms.

accuracy, 19–23, 42, 74
AC motor, 9
angular velocity, 98–99

Banerjee, Shubham, 111–112, **112**
brake, 47, **47**

cam, 26, **27**, 101
Canadarm2, 89–91, **89**
chain drive, 40, 57, 94–95, **96**, 99, **100**
circumference, 44–46, 53, 69, 98
coast, 46, **46**
compound gear, 97, 99
curve turn, 50, **51**

DC motor, 6, 8–12, **8**, **11**, **12**, 17–19, 21–22, 72
degrees of freedom, 62–63, 73, 90
DEKA "Luke" arm, 62
diameter, 44–45, 53
differential gear, 41
differential steering, 40–43, 49–50
driven gear, 94, **94**, 96, **96**, 98
drivetrain, 7, 11, 13, 40, 61, 73, 76, 89, 95, 99
driving gear, 11, 94, **94**, 96–98, **96**

electromagnet, 9, 18, 21–22, 31
encoder, 11–13, 19–21, 72, 77, 99, 105–106, **106**, **108–111**
Engineering Design Process, 67–69, **68**, 87–88

flowchart, 69, 81, 83, **84**, 86, 88
forklift, 6, 24, 56–57, 99
friction, 26, 37–38, 50, 65, 95, 99
fuse, 11–12, 18

gear, 11–12, **11**, **12**, 18–19, 21–25, **28**, 29, 37–38, 40–42, 56–58, 60, **63**, 64, 70–72, **92**, 93–99, **94**, **96**, 101
gear ratio, **94**, 97, 99
grabbing, 7, 26, 28–29, **28**, 55–56, 62–65, **63**, 73, 85, 88, 90, 108, **108–111**

hydraulic, 27, 30

iBot wheelchair, **58**, 59
idler gear, 95, **96**
indent, 101
input, 15–16, 34, 40, 77, 83
International Space Station, **89**, 90–91

LEGO Mindstorms, 9–12, **11**, 17–21, **20**, 24, 26, 28–31, **28**, 33–35, **33**, 41–42, 44–47, **46**, **47**, **49**, **50**, 51, **51**, 56–57, **61**, **64**, 65, 70–72, 80–82, 85, **85**, 99, 111–113, **112**
lever, 25, 40, 51, 54–55, **54**, **55**, 97–98
lifting, 6–8, 23–25, 31–33, 51, 54–57, 71, 73, 99

linear actuator, 22–27, **23**, **27**, **33**, 34–35, 56–57, 60
loop statement, 85–86, 102, 106, **106**, **110–111**

magnetic field, 9, 21, 57
mechanical advantage, 9, 25, 37–40, **39**, 54
miter gear, 101

omniwheel, 38, 40, 48
output, 7, 15, 24–25, 29, 40, 55, 72, 77, 83, 86, 94–95

piezoelectric, 6, 10
pinion, 11, 19, 23–26, 54, 57, 63–64
pivot turn, 50, **50**, 105, **105**, **110**
pneumatic, 6, 10, **14**, 27, 30–35, **32**, **33**, 56–57, 60
point turn, **48**, 49, **49**, 104, **104**
power, 10, 16, 19, 24, 30, 37, 70, 96
power band, 17
precision, 6, 10, 20–23, 26, 74, 105
pseudo-code, 83, 85–86, 108
pulling, 60–61, 96
pushing, 25, 32, 60–61, **61**, 73

rack and pinion, 23–25, 54, 57, 63–64
radius, 25–26, 45, 49, 98

ROBOTC, 44, 78, 82, 85, 101–108, **102**, **104**, **105**, **106**, **107**, **108–111**
roller claw, 65
rotational motor, 11, 15–19, 40, 51, 56, 60, 93
rotational speed, 42, 70

scissor lift, 56
servo motor, 6, 12–13, 19–22, 43, 72, 77, 105, 107, **107**, 108, **108–109**, **111**
simple machines, 8, 22, 38–40
solenoid, 31–32, **32**
solenoid valve, 31, **32**
speed, 12, 16–17, 21–22, 24, 41–44, 49–50, 57, 60, 63, 69–72, 95–98, 102, 104–105
stall, 17–18, 34
statement, 101–102, 106
stepper motor, 9, 21–23, 95
switch, 26, 31, 33–35, **33**, 60
synchro drive, 47–48
system, 7–9, 15, 23–24, 26–27, 30–31, 35, 40–42, 49–50, 56, 60–61, 71–72, 83, 95

3-D printers, 22, 24
torque, 10, 12–13, 16–18, 22–25, 29, 37, 54–55, 57–58, 60, 70–72, 93, 95–98
traction, 37–38, 40, 58, 61, 99

turning, 41, 47–51, **48**, **49**, **50**, **51**, 52–53, 81, 85, **85**, 104–105, **104**, **105**, 108, **110**
turning scrub, 50

VEX, 9, 11–13, **12**, 16–20, **23**, 24, 26, **27**, 30–31, 41, 43–44, 48, 51, 57, 60, **63**, 65, 72, 78, 97, 99, 101–108, **102**, **103**, **104**, **105**, **106**, **107**, **108–111**, 111
vise grip, 57, 64–65

wedge, 39, **39**
wheelbase, **52**, 53
worm gear, 22–24, **23**, **28**, 29, 56–57, 64

About the Author

The Pereiras are a mother/daughter team. Their love of Isaac Asimov's stories created a passionate interest in robots, which they developed further with their introduction to LEGO Mindstorms in Girl Scouts. Lana has taught Middle School Robotics for several years, as well as coaching FIRST LEGO League teams and mentoring a VEX team. Lidia, a sculptor, was a Girl Scout STEM trainer and VEX robotics coach, and remains a FIRST Lego League mentor.